FIBER OPTICS
through
EXPERIMENTS

FIBER OPTICS *through* EXPERIMENTS

Third Edition

Edited by

M.R. Shenoy
Department of Physics, IIT Delhi

Sunil K. Khijwania
Department of Physics, IIT Guwahati

Ajoy Ghatak
Department of Physics, IIT Delhi

&

Bishnu P. Pal
Department of Physics, IIT Delhi

Viva Books

New Delhi | Mumbai | Chennai | Kolkata | Bengaluru | Hyderabad | Kochi | Guwahati

Publisher's note

Every possible effort has been made to ensure that the information contained in this book is accurate at the time of going to press, and the publisher and author cannot accept responsibility for any errors or omissions, however caused. No responsibility for loss or damage occasioned to any person acting, or refraining from action, as a result of the material in this publication can be accepted by the editor, the publisher or the author.

Every effort has been made to trace the owners of copyright material used in this book. The author and the publisher will be grateful for any omission brought to their notice for acknowledgement in the future editions of the book.

First Published 1994
Second Edition 2009
Third Edition 2011
Reprinted 2013, 2015, 2017, 2018, 2024

Viva Books Private Limited

- 4737/23, Ansari Road, Daryaganj, New Delhi 110 002
  Tel. 42242200, 23258325, 23283121, Email: vivadelhi@vivagroupindia.net
- 76, Service Industries, Shirvane, Sector 1, Nerul, Navi Mumbai 400 706
  Tel. 27721273, 27721274, Email: vivamumbai@vivagroupindia.net
- Megh Towers, Old No. 307, New No. 165, Poonamallee High Road, Maduravoyal
  Chennai 600 095 Tel. 23780991, 23780992, Email: vivachennai@vivagroupindia.net
- B-103, Jindal Towers, 21/1A/3 Darga Road, Kolkata 700 017
  Tel. 22816713, Email: vivakolkata@vivagroupindia.net
- 7, GF, Sovereign Park Aptts., 56-58, K. R. Road, Basavanagudi, Bengaluru 560 004
  Tel. 26607409, Email: vivabangalore@vivagroupindia.net
- 101-102, Moghal Marc Apartments, 3-4-637 to 641, Narayanguda, Hyderabad 500 029
  Tel. 27564481, Email: vivahyderabad@vivagroupindia.net
- First Floor, Beevi Towers, SRM Road, Kaloor, Kochi 682 018
  Tel. 0484-2403055, 2403056, Email: vivakochi@vivagroupindia.net
- 232, GNB Road, Beside UCO Bank, Silpukhuri, Guwahati 781 003
  Tel. 0361-2666386, Email: vivaguwahati@vivagroupindia.net

www.vivagroupindia.com

ISBN: 978-81-309-1811-2

Published by Vinod Vasishtha for Viva Books Private Limited, 4737/23 Ansari Road, Daryaganj, New Delhi 110 002.

Printed and bound at Replika Press, Kundli, Haryana.

Contents

PART III

Preface

The invention of laser in 1960 and the fabrication of low-loss optical fibers in 1970 have led to an immense growth in research and development in the general area of Photonics. In particular, fiber optics has revolutionized the communication industry as the most attractive and viable transmission medium for wideband telecommunication, data transmission and information transfer. This has also led to an explosive growth and emergence of many new applications of fiber optics. These include a variety of fiber optic sensors, in-line guided wave components like polarizers, directional couplers, fiber gratings, fiber amplifiers, optical switches, etc. for sensing, instrumentation and optical signal processing. In addition to its tremendous technological importance, fiber optics also provides us with an excellent opportunity to demonstrate and elucidate a large number of physical concepts. Because of all these, fiber optics is slowly but surely entering the undergraduate and postgraduate curricula in many universities, both in science and engineering.

For over 25 years, we have been teaching fiber optics at the undergraduate as well as at the M.Sc and M.Tech levels at IIT Delhi. During this period, a large number of simple experiments have evolved, which can be carried out as a part of the laboratory training for students, both at the B.Sc/B.Tech and the M.Sc/M.Tech levels. These experiments have also formed a part of the training of participants in various summer and winter schools on fiber optics and related areas that we have conducted at IIT Delhi under continuing education program of the Institute. More recently, we have also been involved in developing a Fiber Optic Education kit, which is being manufactured by YOLUX in Bhubaneshwar, Orissa. Several experimental results, which are mentioned in the book as sample results within certain chapters, were based on experiments performed with this kit. Since a number of interesting experiments on fiber optics can be carried out by students at the UG/PG level, we felt that a book describing the details of few such experiments could be useful to teachers in Physics and Electrical Engineering, who might be interested in setting up a laboratory on fiber optics in their institutions. It is with this motivation that we have written this book. Emphasis of the contents is mainly on the characterization of optical fibers for telecommunication and sensor applications; organization of the contents is described in Chapter 1. Any suggestions for improvement of the book would be greatly appreciated. In particular, we will be most grateful if errors are pointed out to us; our email addresses are: mrshenoy@gmail.com, skhijwania@yahoo.com, ajoykghatak@gmail.com and bishnupal@gmail.com.

The first edition of this book came out in 1994. We have added a few chapters covering important topics like bend induced loss in optical fibers, Fiber Bragg Gratings, etc. and also revised quite a few chapters.

We are very grateful to Dr. C.S. Goh (Alnair Labs, Japan), and our colleagues at IIT Delhi Professor I.C. Goyal, Professor B.D. Gupta, Professor Arun Kumar, Mr. Varghese Paulose, Professor Anurag Sharma, Dr Ramanand Tewari, Professor K. Thyagarajan, and Dr. R.K. Varshney, for agreeing to contribute chapters to this book; Mr. Varghese Paulose is now with Institute for Infocomm Research in Singapore and Dr. Ramanand Tewari is now with RSOFT in Ottawa, Canada. Unfortunately, Professor I.C. Goyal is no more with us – we miss him greatly. All the above-mentioned colleagues of ours at IIT Delhi have also made significant contributions in the development of the Fiber Optics Laboratory at IIT Delhi since its inception in late 1970s. This laboratory grew over the years with generous support received through projects sponsored by Government of India departments like MHRD, DST, IFCPAR/CEFIPRA (Society of DST), DoE (now DIT), CSIR, and overseas government funding agencies like Overseas Development Association (ODA, UK), Office of Science and Technology (UK), and NIST (Boulder, USA). IIT Delhi administration provided all the necessary infrastructure support for its growth.

One of us (AG) is grateful to Department of Science and Technology, Government of India for support; a part of this revised edition was carried out with partial support from DST.

M.R. Shenoy
Sunil K. Khijwania
Ajoy Ghatak
Bishnu P. Pal

July 2008

1 General Introduction

A general *optical fiber system* comprises of an optical transmitter, a fiber link, (which forms the transmission medium), and a receiver (see Fig. 1.1). The transmitter consists of a light source, which may be modulated by a suitable drive circuit in accordance with the signal to be transmitted. Similarly, the receiver consists of a detector, followed by an electronic amplifier and a signal recovery unit. In many systems, such as optical fiber sensors, the transmitter may consist only of a stable light source, and the signal to be detected at the receiver may be generated enroute inside the fiber. The receiver may be a simple power meter, which consists of a photodetector followed by an amplifier and a display unit. Thus, the basic optical fiber system would consist of a source and a photodetector at the two ends of the fiber link.

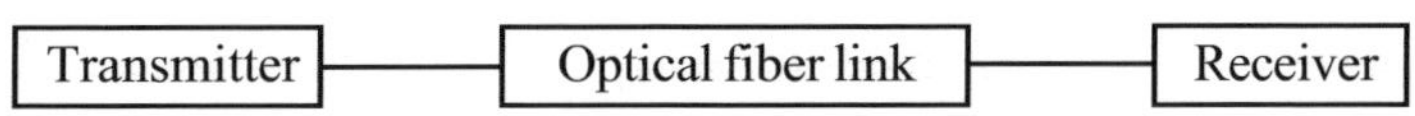

Fig. 1.1: Schematic of a general optical fiber system

1.1 Light sources

The source may be a light emitting diode (LED), a laser, or a broadband source, such as a tungsten halogen lamp. Choice of the source depends on the application. For example, in a laboratory demonstration experiment, a laser emitting in the visible region such as a simple He-Ne laser may be the preferred source, while in commercial applications a compact and rugged semiconductor source such as a laser diode (LD) or an LED would be the preferred source. In any case, it is very essential to have knowledge of the source characteristics, both electrical and optical, before using the source for a particular application. The maximum optical power output, wavelength of emission, linewidth, and radiation pattern and modulation bandwidth are some of the basic optical characteristics that one needs to know. The electrical connections, operating voltage, current and their maximum ratings must be known precisely before switching on the source. Special precautions must be taken in operating laser diodes; a proper drive circuit and a stable, surge-free power supply is necessary in this case. Heat sink and cooling arrangements would be necessary for high-power sources. Laser safety is an important aspect, which the user should be aware of; safety goggles/protective wears must be used with high-power lasers. Most of the experiments described in this book, however, are based on He-Ne lasers of a few milliwatt power output; safety goggles are not very essential in this case, provided one takes certain basic precautions while mounting, aligning and coupling of the source to the fiber.

1.2 Photodetectors

A PIN-photodiode is the most common photodetector used in optical fiber systems. However, depending on the application, for example, to detect extremely weak signals, an avalanche photodiode (APD) or a photomultiplier tube (PMT) may be used. Again, the detector characteristics such as detector sensitivity, active area, linearity, bandwidth, noise equivalent power (NEP), and wavelength range of operation, electrical biasing and the maximum electrical ratings should be known. For example, silicon detectors are widely used in the 0.4–1.0 μm wavelength range, while germanium and InGaAs detectors are used for measurements in the 0.9–1.7 μm range. Large area detectors are very useful in many applications, while high-speed (pulse) detection normally requires small area detectors in order to minimize detector capacitance. Detection of very low-level signals often requires lock-in-amplification to increase the signal to noise ratio (SNR). Some applications may require cooling the detector on a Peltier element, or even in liquid nitrogen ambience.

In several experiments described in this book, it is required to measure the light intensity distribution across the fiber output. To do this, one can use either a pinhole masked large area detector, or just a small area detector; in the latter case, the diameter (or the width) of the active area determines the spatial resolution of the measurement. The pinhole masked large area detector has the advantage of using different sizes of pinholes, depending on the requirement of spatial resolution of the measurement. Henceforth, we shall refer to both the above schemes as 'pinhole detector'.

1.3 Fiber link

The transmission medium, viz. the optical fiber, forms the link between the transmitter and the receiver, and its characteristics are also of vital importance to the performance of the system.

Depending on the application, *single-mode fibers* or *multimode fibers* may be used. As the names indicate, a single-mode fiber supports only one guided mode, i.e. at the operating wavelength, only one specific field distribution can propagate along the fiber with very little attenuation. On the other hand, multimode fibers support a large number of guided modes. Today, almost all long-haul, high-speed communication links are based on single-mode fiber designs which exhibit very low *loss* and *dispersion* in the lowest loss window of silica fibers. Cutoff wavelength, mode field diameter (MFD), and attenuation coefficient are some of the important characteristics of a single-mode fiber. All of these can be determined individually by performing several experiments as illustrated in this book. Alternatively, if one can precisely measure the refractive index profile of the fiber, almost all characteristics of the fiber (single-mode or multimode) can be estimated. However, it is very difficult to precisely measure the refractive index profile of a single-mode fiber, and one requires very sophisticated equipments or setups to measure the same. Even then, the measured profiles may not be of specific forms that have analytical solutions to the wave equation, and may lead to further difficulty in extracting the various characteristic parameters. In this book, we have described a few simple experiments to determine some of the important characteristics of single-mode fibers, including highly birefringent fibers.

Multimode fibers are much easier to work with due to their large core diameter. While 50 μm is a typical (standard) value for the core diameter of a multimode fiber, specific applications may require fibers with much larger core diameters. Presently, multimode fibers are mainly used in applications such as short distance communication links, fiber sensors and as optical power transporting medium. It may be mentioned here that a variety of fibers such as plastic fibers, plastic-clad silica fibers, and fibers with large numerical aperture (NA) have been developed by workers in the area of optical fiber sensors for specific applications. For multimode fibers, measurement of fiber characteristics, such as numerical aperture, refractive index profile, and spectral attenuation can be done conveniently in the laboratory.

Single-mode fibers can be thought of as a special case of multimode fibers wherein a single propagating mode exists. It is normally not possible, by merely looking at the fibers, to differentiate between a standard single-mode fiber and a standard multimode fiber since both have the same total (core + cladding) diameter of 125 μm. It becomes simple, however, to identify the fiber if one couples light from a He-Ne laser into the given fiber. If the fiber is (highly) multimoded, then the fiber output (projected on a paper-screen) will comprise of a speckle pattern, due to interference among the various modes; even the slightest disturbance caused to the fiber, anywhere along its length, would result in an observable change in the speckle pattern. In the case of a single-mode fiber (at the He-Ne wavelength), the output will consist of a clear spot with radially uniform intensity distribution, varying smoothly from a maximum at the centre of the spot to the background illumination towards the periphery. In the case of a few-mode fiber (supporting about 2–6 modes), depending on the input excitation, intensity lobe corresponding to a higher order mode or a superposition of modes can be seen at the output. For example, a single-mode fiber designed for operation at $\lambda \approx 1.3$ μm becomes a few moded fiber at the He-Ne wavelength ($\lambda = 0.6328$ μm). It is worthwhile to try and selectively excite the higher order modes and observe the corresponding intensity pattern at the output.

1.4 Light coupling

After an introductory discussion about the three basic components of an optical fiber system, we need to consider the important aspect of coupling of these components. Light from the source to the fiber may be coupled by direct butting (called as *butt coupling*), or by using a collimating-and-focusing lens system. Butt coupling cannot be employed for sources emitting over a large surface area; butt coupling can be efficient when the size of the source is nearly equal to the size of the fiber core. For example, butt coupling is often employed in the fabrication of fiber-pigtailed semiconductor sources (LED and LD); the fibers in this case are usually multimode fibers. Use of a graded index (GRIN) lens or a pair of suitably chosen lenses is normally required to achieve efficient coupling of light from a laser diode into a single-mode fiber. The details of this coupling, though important, will not be discussed in this book since the experiments described in this book do not require the use of a laser diode as the source.

Use of a microscope objective (MO) is a convenient means of coupling light from a source into a fiber in the laboratory. Normal MOs typically comprise of a combination of three lenses, while objectives with special imaging requirements may have a number of lenses forming a complex lens system. One usually finds the value of two parameters, namely 'magnification' and 'numerical aperture', inscribed on the metallic case of the objective. Depending on the situation, one has to choose an appropriate MO for light coupling. Accordingly, in the experiments described in this book, MOs of different magnification have been recommended at different places. The details of launching light into a fiber (and decoupling light from the exit-end of the fiber) is discussed in Chapter 5.

1.5 Organization of the book

In this book, we are primarily concerned with the characteristics of the transmission medium, viz., the optical fiber. Several simple-to-do experiments using both single-mode and multimode fibers have been described. The contents of the book are divided into three parts. In Part I, a review of the basic theory is provided through Chapters 2–4, which should form as an introduction to the beginners, as well as a ready reference material to those who are familiar with the subject. Experiments have been clubbed into two groups, which we call 'Level I'and 'Level II', and these form the contents of Part II and Part III. Usage of the word 'Level' in grouping the experiments is perhaps bit of a misnomer, in the sense that they really do not distinguish experiments of two different levels of depth and understanding. The basic criterion applied in distinguishing the two groups is the level of sophistication/sensitivity required in the detection part of the measurements. For example, to determine the V-number of a single-mode fiber from a measurement of its far-field, one needs a very good detector—power-meter arrangement with a dynamic range in excess of 50 dB, and a careful scan of the far-field is necessary; while the mode field diameter of the fiber can be determined by using a simple detection system with dynamic range $\approx$25 dB. Accordingly, the above two experiments, though very similar, are arranged in two different groups. Thus, almost all experiments in the first group can be performed with a simple photodetector-amplifier combination using a low power He-Ne laser (2 mW) and a tungsten halogen lamp as the source. However, some of the experiments in the second group would require a relatively high-power source and also some special purpose equipments/components, in addition to the requirements on the detection system (mentioned above).

As mentioned in the beginning of this introduction, a light source and a photodetector are the two other basic components of an optical fiber system, and it is very essential to know the characteristics of the source and the detector as well. Again, there can be a number of experiments to characterize different types of sources and detectors, and to characterize the larger units of transmitter and receiver. The latter part deals with the different types of signals, coding, modulation, demodulation, receiver sensitivity, etc. and forms the basics of communication engineering. Students in Optoelectronics/Optical Communication are required to have knowledge of these aspects, and hence some of these experiments are included in the third and final part of this book. For example, in view of the importance of the source and detector characteristics in an optical fiber system, a set of experiments to characterize a typical light source like LED are described in detail. A simple display LED has been chosen to illustrate the procedure for characterization; some modifications may be necessary for specific sources. Finally, two experiments to characterize fiber Bragg gratings (FBG), which are finding very important applications in communication and sensing, have been presented.

PART I

2

Basic Optics

Ajoy Ghatak

2.1 Introduction

In this chapter we will briefly discuss the phenomena of polarization, interference and diffraction of a light wave. The concepts developed in this chapter would form the basic building blocks for the remaining part of the book. For more details on these basic concepts, the reader may refer to the books listed at the end of this chapter.

2.2 Plane polarized beams

A linearly polarized plane wave is the simplest electromagnetic wave, and if we assume the plane wave to be propagating in the $+z$ direction, the corresponding electric field can be written in the form

$$\mathbf{E} = \hat{\mathbf{x}} E_0 \cos(\omega t - kz + \theta) \tag{2.1}$$

where the electric field vector is assumed to oscillate along the x direction. In Eq. (2.1), E_0 is the amplitude of the wave, $\omega\,(= 2\pi\nu)$ is the angular frequency, θ is a constant phase term, and k is the propagation constant. For propagation in free space

$$k = \frac{\omega}{c} \tag{2.2}$$

where c represents the velocity of light in free space; when propagating in a medium characterized by refractive index n, we have

$$k = \frac{\omega}{v} \tag{2.3}$$

where
$$v = \frac{c}{n(\omega)}$$

represents the velocity of the electromagnetic wave in that medium. Equation (2.1) describes an x-polarized wave. It is also known as a *linearly polarized* wave because the electric vector is

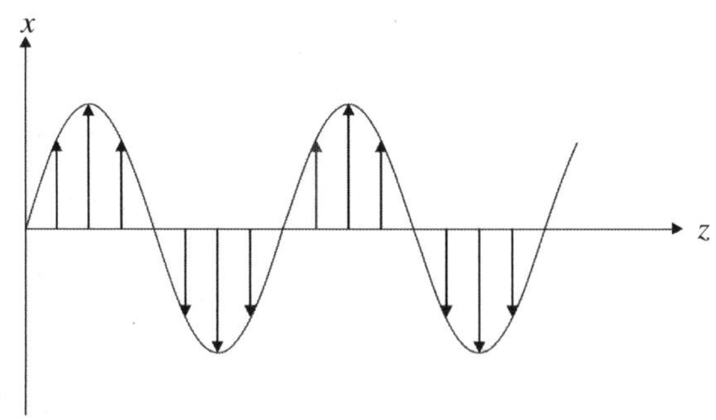

Fig. 2.1: An x-polarized plane electromagnetic wave propagating along the z direction. The arrows represent the direction and magnitude of the electric field vector at a particular instant of time.

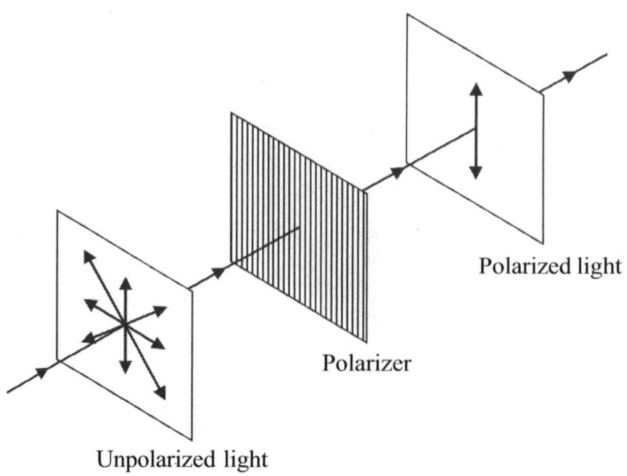

Fig. 2.2: Production of a plane polarized light from an unpolarized light source using a linear polarizer. The vertical lines in the polarizer represent the pass axis.

oscillating along a specific axis (see Fig. 2.1). A linearly polarized wave is also alternatively known as a *plane polarized* wave because its electric field is always confined to a particular plane; for the x-polarized wave propagating along the z direction, the electric field is confined to the x-z plane (see Fig. 2.1). Note that the plane waves are transverse in nature, i.e. $E_z = 0$.

2.2.1 Production of plane polarized beams

The ordinary light beam, like the one coming from a sodium lamp or from the sun, is unpolarized, because its electric vector (on a plane transverse to the direction of propagation) keeps changing its direction in a random manner as shown in Fig. 2.2. If we allow the unpolarized beam to fall on a piece of Polaroid sheet then the beam emerging from the Polaroid will be linearly polarized. In Fig. 2.2 the vertical lines shown on the Polaroid represent what is usually referred to as the "pass-axis" of the Polaroid; i.e. the Polaroid absorbs the electric field perpendicular to its pass-axis. Polaroid sheets are extensively used for producing linearly polarized light beams. As an interesting corollary, we may note that if a second Polaroid (whose pass-axis is at right angles to the pass-axis of the first Polaroid) is placed immediately after the first Polaroid, then no light will come through it; the Polaroid's are said to be in a "crossed-position".

2.3 Circularly and elliptically polarized beams

Let us consider the superposition of two plane waves, one polarized in the x direction and the other polarized in the y direction, with a phase difference of $\pi/2$ between them:

$$\mathbf{E}_1 = E_0\hat{\mathbf{x}}\cos(\omega t - kz) \tag{2.4}$$

$$\mathbf{E}_2 = E_0\hat{\mathbf{y}}\cos\left(\omega t - kz - \frac{\pi}{2}\right) \tag{2.5}$$

where we have assumed the amplitude of the two beams to be the same ($= E_0$). The resultant electric field is given by

$$\mathbf{E} = E_0[\hat{\mathbf{x}}\cos(\omega t - kz) + \hat{\mathbf{y}}\sin(\omega t - kz)] \tag{2.6}$$

which describes a *right circularly polarized* (RCP) wave. At any particular value of z, the tip of the E-vector, with increasing time t, can easily be shown to rotate on the circumference of a circle. For example, at $z = 0$ the x and y components of the electric vector are given by

$$E_x = E_0 \cos \omega t, \quad E_y = E_0 \sin\omega t; \tag{2.7}$$

thus the tip of the electric vector rotates on a circle in the clockwise direction (see Fig. 2.3) and therefore it is said to represent a RCP beam. When propagating in air or in any isotropic medium,

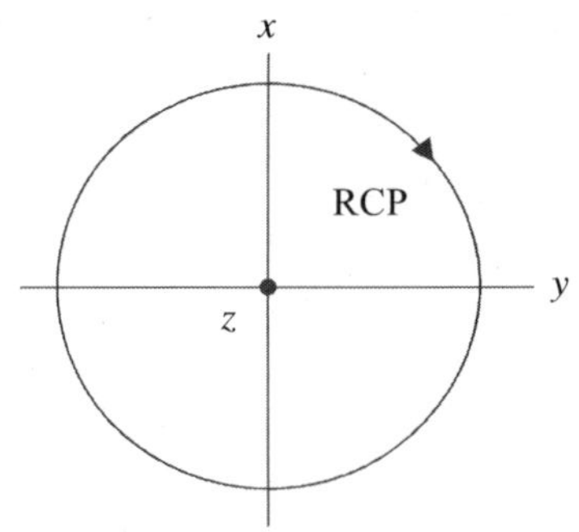

Fig. 2.3: Representation of a right circularly polarized (RCP) beam propagating along the z direction which is into the page.

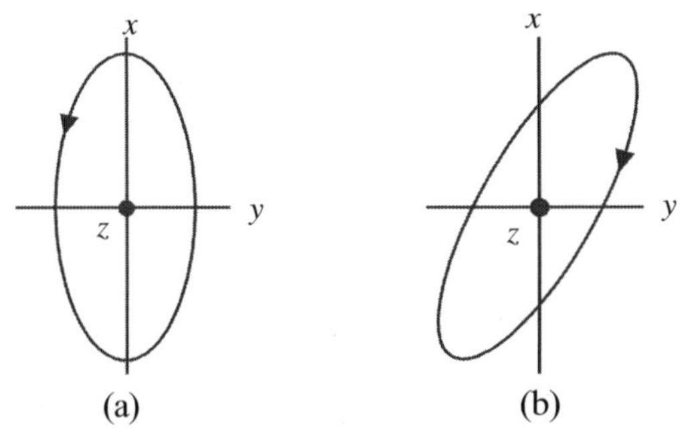

Fig. 2.4: Two different elliptical states of polarization represented by Eq. (2.8) with

(a) $E_1 = \frac{E_0}{2}$ and $\phi = -\frac{\pi}{2}$, and (b) $E_1 = E_0$ and $\phi = \frac{\pi}{3}$.

the state of polarization (SOP) is maintained, i.e. a linearly polarized beam will remain linearly polarized; similarly, a RCP beam will remain RCP. In general, the superposition of two beams with arbitrary amplitudes and phase

$$E_x = E_0 \cos(\omega t - kz) \text{ and } E_y = E_1 \cos(\omega t - kz - \phi) \tag{2.8}$$

will represent an elliptically polarized beam. Figure 2.4 shows two different elliptically polarized beams. Figure 2.4(a) corresponds to

$$E_1 = \frac{1}{2} E_0 \text{ and } \phi = -\frac{\pi}{2} \tag{2.9}$$

and Fig. 2.4(b) corresponds to

$$E_1 = E_0 \text{ and } \phi = +\frac{\pi}{3} \tag{2.10}$$

2.4 Propagation of a light beam through a quarter wave plate (QWP)

A quarter wave plate is a device, which is used in many experiments involved in fiber and integrated optics, and is basically used to change the SOP of a propagating optical beam. For example, using this device (as we will see) we may transform a circularly polarized beam to a linearly polarized beam, and vice versa. A QWP is made of an anisotropic medium like calcite or quartz. We will not go into the details of an anisotropic medium; it will suffice here to say that inside a crystal like that of calcite, quartz, $LiNbO_3$, etc. there is a preferred direction known as the *optic axis*. In a QWP the crystal is cut in such a way that the optic axis is at right angles to the thickness of the plate. Figure 2.5 shows a QWP with its optic axis along the y direction. The "modes" of the QWP are x- and y-polarized; the y-polarized mode, which is polarized along the optic axis is usually referred to as the extraordinary wave, and the x-polarized mode as the ordinary wave. By "modes" we imply that if the incident wave is x-polarized, then it will propagate as an x-polarized wave; similarly, if the incident wave is y-polarized, then it will propagate as a y-polarized wave without any change in the SOP, but with a velocity slightly different than the velocity of the x-polarized wave. On the other hand, if a circularly polarized beam is incident, then we must express this as a linear combination of two x- and y-polarized waves, and then consider the independent propagation of the two beams.

As an example, we consider the incidence of a beam polarized at an angle of 45° to the x-axis (see Fig. 2.5). At $z = 0$, it can be expressed as

$$E_x = E_0 \sin\theta \cos\omega t; \; E_y = E_0 \cos\theta \cos\omega t, \tag{2.11}$$

Crystal	n_o	n_e	@ λ (μm)
Calcite	1.658	1.486	0.589
Quartz	1.544	1.553	0.589
Rutile	2.621	2.919	0.579

where $\theta = 45°$. The refractive indices "seen" by the x- and y-polarized waves are different, and are usually denoted by n_o and n_e, respectively. These are known as the ordinary and extraordinary refractive indices, and are known constants of the crystal (see the above table). Thus, the x- and y-polarized waves propagate with velocities c/n_o and c/n_e respectively, and therefore inside the crystal, we will have

$$E_x = \frac{E_0}{\sqrt{2}} \cos(\omega t - k_o z) \tag{2.12}$$

and

$$E_y = \frac{E_0}{\sqrt{2}} \cos(\omega t - k_e z) \tag{2.13}$$

where

$$k_o = \frac{\omega}{c} n_o \text{ and } k_e = \frac{\omega}{c} n_e \tag{2.14}$$

Thus, as the beam propagates, a phase difference is set-up between the two components. We assume $n_o > n_e$ (which is true for negative crystals like calcite). Let the thickness d of the crystal be such that

$$(k_o - k_e)d = \frac{\pi}{2}$$

or

$$d = \frac{\lambda}{4(n_o - n_e)} \tag{2.15}$$

where λ (= $2\pi c/\omega$) represents the free space wavelength. Then at $z = d$, the x and y components of the polarized waves are given by

$$E_x = \frac{E_0}{\sqrt{2}} \cos(\omega t - \phi) \tag{2.16}$$

and

$$E_y = \frac{E_0}{\sqrt{2}} \cos\left[\omega t - \phi + \frac{\pi}{2}\right]$$

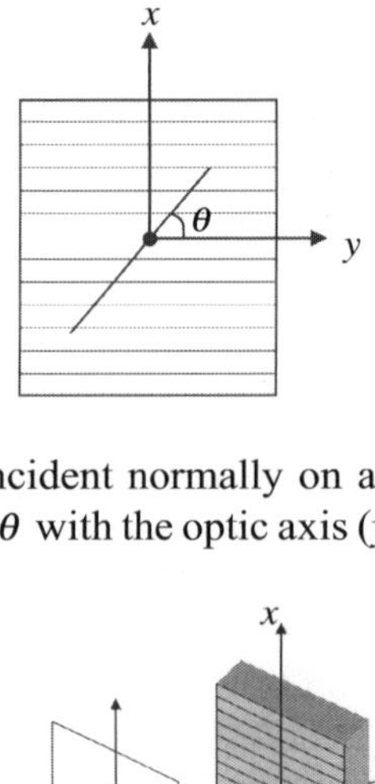

Fig. 2.5: A plane polarized beam is incident normally on a quarter wave plate with the plane of polarization making an angle θ with the optic axis (y-axis in the figure).

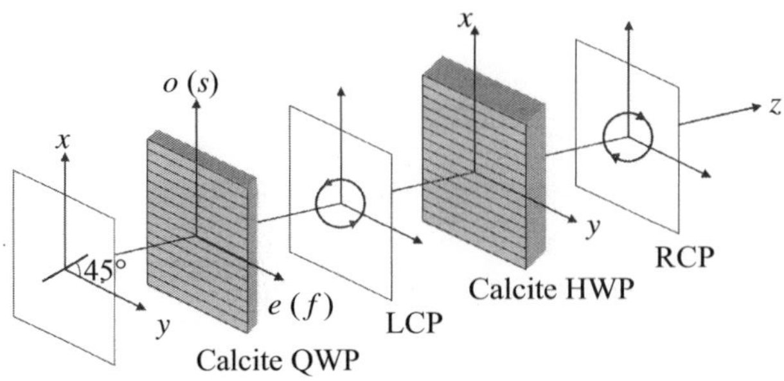

Fig. 2.6: A linearly polarized beam making an angle 45° with the y-axis gets converted to a LCP after propagating through a calcite QWP; further, an LCP gets converted to a RCP after propagating through a calcite HWP. The optic axis is here assumed to be along the y-axis. Figure adapted from Ajoy Ghatak's *Optics*, 4th Edition, McGraw-Hill, New Delhi (2008).

or,
$$E_y = -\frac{E_0}{\sqrt{2}} \sin(\omega t - \phi) \tag{2.17}$$

where $\phi = k_o d$. Equations (2.16) and (2.17) describe a *left circularly polarized* (LCP) beam.

A *half-wave plate* (HWP), similarly, is another device, which introduces a phase difference of π between the two electric field components; thus its thickness will be twice the thickness of a QWP. A linearly polarized light beam remains linearly polarized after passing through a HWP; however, the orientation of the plane of polarization can be changed by a desired amount using a HWP.

As a simple example, Fig. 2.6 shows a linearly polarized beam making an angle 45° with the y-axis gets converted to a LCP after propagating through a calcite QWP; the y-axis is the direction of the optic axis of the QWP. Further, the LCP gets converted to a RCP after propagating through a calcite HWP.

2.5 Two beam interference

Whenever two waves superpose, one obtains an intensity distribution, which is known as the interference pattern. We will consider here the interference pattern produced by waves emanating from two point sources. As is well known, a stationary interference pattern is observed when the two interfering waves maintain a constant phase difference. For light waves, due to the very process of emission, one cannot observe a stationary interference pattern between the waves emanating from two independent sources[1] although interference does take place. Thus, one tries to derive the interfering waves from a single wave so that a definite phase relationship is maintained all through.

Let S_1 and S_2 represent two coherent point sources emitting waves of wavelength λ [see Fig. 2.7(a)]. We wish to determine the interference pattern on the photographic plates P_1 and P_2. The intensity distribution is given by:

$$I = 4I_0 \cos^2 \frac{\delta}{2} \tag{2.18}$$

[1] It is difficult to observe the interference pattern even with two laser beams [see, e.g. Chapter 17 of Ref. 1].

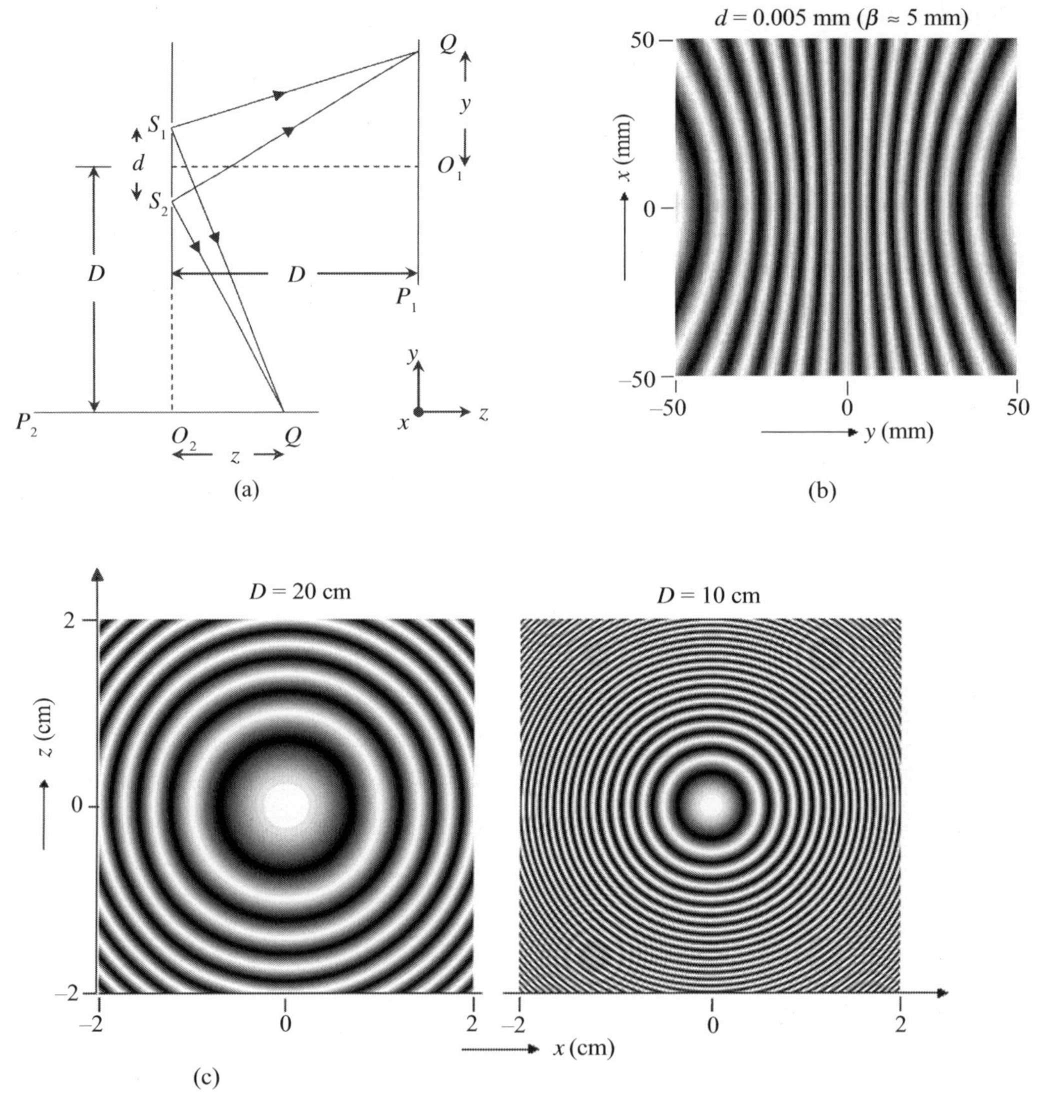

Fig. 2.7: (a) Schematic of the arrangement to observe two-beam interference; (b) fringe pattern on the plane P_1 and (c) fringe pattern on the plane P_2. Computer generated interference patterns are adapted from Ajoy Ghatak's *Optics*, 4th Edition, McGraw-Hill, New Delhi (2008).

where I_0 is the intensity produced by either of the waves independently and

$$\delta = \frac{2\pi}{\lambda}\Delta \tag{2.19}$$

where

$$\Delta = S_1Q - S_2Q \tag{2.20}$$

represents the path difference between the two interfering waves. Thus, when

$$\delta = 2n\pi \Rightarrow \Delta = S_1Q - S_2Q = n\lambda; \quad n = 0, 1, 2, \ldots\ldots \text{ (Bright Fringe)} \tag{2.21}$$

we will have a bright fringe, and when

$$\delta = (2n+1)\pi \Rightarrow \Delta = S_1Q - S_2Q = \left(n + \frac{1}{2}\right)\lambda; \quad n = 0, 1, 2, \ldots\ldots \text{ (Dark Fringe)} \tag{2.22}$$

we will have a dark fringe. Using simple geometry one can show that the locus of the points (on the plane P_1) such that $S_1Q \approx S_2Q = D$ is a hyperbola, given by

$$\left(d^2 - \Delta^2\right)y^2 - \Delta^2 x^2 = \Delta^2\left[D^2 + \frac{1}{4}\left(d^2 - \Delta^2\right)\right] \tag{2.23}$$

Now, $\Delta = 0 \Rightarrow y = 0$

which represents the central bright fringe. Equation (2.23) can be written in the form [1]

$$y = \sqrt{\frac{\Delta^2}{d^2 - \Delta^2}}\left[x^2 + D^2 + \frac{1}{4}\left(d^2 - \Delta^2\right)\right]^{\frac{1}{2}} \tag{2.24}$$

For values of y such that $x^2 << D^2$ (2.25)

the loci are straight lines parallel to the x-axis and one obtains straight line fringes as shown in Fig. 2.7(b). The corresponding fringe width would be

$$\beta = \frac{\lambda D}{d} \tag{2.26}$$

Thus, for $D = 50$ cm, $d = 0.05$ cm and $\lambda = 6000$ Å, we get $\beta = 0.06$ cm.

We next consider the interference pattern produced on the plane P_2. As an example, let us assume for simplicity $\lambda = 6 \times 10^{-5}$ cm and $d = 6 \times 10^{-3}$ cm, so that d is an integral multiple of λ. Thus

$$S_1O_2 - S_2O_2 = d = 100\lambda = p\lambda$$

and the point O_2 will be a bright spot. Obviously, the fringe pattern on the plane P_2 will be circular [see Fig. 2.7(c)] because the locus of the point Q such that $S_1Q - S_2Q$ is a constant will be a circle on the x-z plane whose center is at O_2. The point Q will be bright if

$$S_1Q - S_2Q = (S_1O_2 - S_2O_2) - n\lambda = (p - n)\lambda$$

where $n = 0, 1, 2, \ldots\ldots$. Elementary algebra shows that the radius z of the fringes are given by

$$z = \left[\left\{\frac{Dd}{(p-n)\lambda} - \frac{1}{2}(p-n)\lambda\right\}^2 - \left(D - \frac{1}{2}d\right)^2\right]^{\frac{1}{2}} \tag{2.27}$$

and it will correspond to the $(p-n)^{th}$ bright fringe. For example, for $d = 0.6$ mm, $\lambda = 6 \times 10^{-5}$ cm, $D = 50$ cm, we have $p = 1000$, and the fringes corresponding to $n = 1$, 2 and 3 (which will correspond to the 999th, 998th, 997th fringe) will have radii 2.24 cm, 3.17 cm and 3.88 cm, respectively.

It can easily be seen that if the distance d is increased, the central fringe will become a higher order fringe and thus the fringes will emerge from the center. On the other hand, if the distance d is decreased the central fringe will become a lower order fringe and thus the fringes will collapse to the center. We may mention here that using single-mode fibers, one can easily set-up a Mach-Zehnder interferometer arrangement to produce such circular fringes as discussed in Chapter 20.

2.6 Diffraction of a Gaussian beam

The diffraction phenomenon can be easily understood by considering the diffraction divergence of a Gaussian beam. Indeed, when a laser oscillates in its fundamental mode, the transverse amplitude distribution is Gaussian; similarly, the transverse amplitude distribution of the fundamental mode of an optical fiber is very nearly Gaussian. Therefore the study of the diffraction of a Gaussian beam is of considerable importance in fiber optics.

We consider a Gaussian beam propagating along the z direction whose amplitude distribution on the plane $z = 0$ is given by

$$u(x, y, 0) = A \exp\left[-\frac{x^2 + y^2}{w_0^2}\right] \tag{2.28}$$

where w_0 is usually referred to as the "spot size" of the beam; it represents the radial distance at which the intensity falls off by a factor of e^2. The above equation implies that the phase front is plane at $z = 0$. As the beam propagates along the z-axis, diffraction occurs and one obtains

$$u(x, y, z) = \frac{iA\pi}{\lambda}\frac{2w_0^2}{2z + ikw_0^2}\exp\left[-ik\left(z + \frac{x^2 + y^2}{2R(z)}\right)\right]\exp\left[-\frac{x^2 + y^2}{w^2(z)}\right] \tag{2.29}$$

where

$$R(z) = z\left[1 + \frac{\pi^2 w_0^4}{\lambda^2 z^2}\right] \tag{2.30}$$

represents the radius of curvature of the wavefront and

$$w(z) = w_0\left(1 + \frac{\lambda^2 z^2}{\pi^2 w_0^4}\right)^{1/2} \tag{2.31}$$

represents the z-dependent spot size of the beam. For large values of z,

$$w(z) = \frac{\lambda z}{\pi w_0} \tag{2.32}$$

implying

$$\tan\theta \approx \frac{w(z)}{z} \approx \frac{\lambda}{\pi w_0} \tag{2.33}$$

where θ is the semi-angle of the cone defining the diffraction divergence of the beam. Thus, for $\lambda \approx 0.6\ \mu m$,

$$\theta \approx 0.011^0 \text{ for } w_0 = 1 \text{ mm} \Rightarrow w \approx 2.6 \text{ mm at } z = 10 \text{ m}$$

and

$$\theta \approx 1.09^0 \text{ for } w_0 = 10\ \mu m \Rightarrow w \approx 39.5 \text{ mm at } z = 2\ \mu m$$

(see Fig. 2.8). Notice that θ increases with decrease in w_0 (smaller the size of the aperture, greater is the diffraction). Further, for a given w_0, θ decreases with decrease in λ; indeed, as $\lambda \to 0$ there is no diffraction, which is the geometrical optics limit.

2.7 Intensity distribution at the back focal plane of a lens

If a truncated plane wave of diameter $2a$ propagating along the z-axis is incident on a converging lens of focal length f [see Fig. 2.9(a)], the intensity distribution on the front focal plane is given by [see, e.g. Ref. 2]

$$I = I_0\left[\frac{2J_1(v)}{v}\right]^2 \tag{2.34}$$

where

$$v = \frac{2\pi a}{\lambda f} r,$$

I_0 is the intensity at the axial point F and r is the distance from the point F on the focal plane. Equation (2.33) describes the well-known Airy pattern [see Fig. 2.9(b)]. About 84% of the light energy is contained within the first dark ring and about 7% of light energy is contained in the annular region between the first two dark rings, etc. the first two dark rings occurring at

$$v = 3.832 \text{ and } 7.016$$

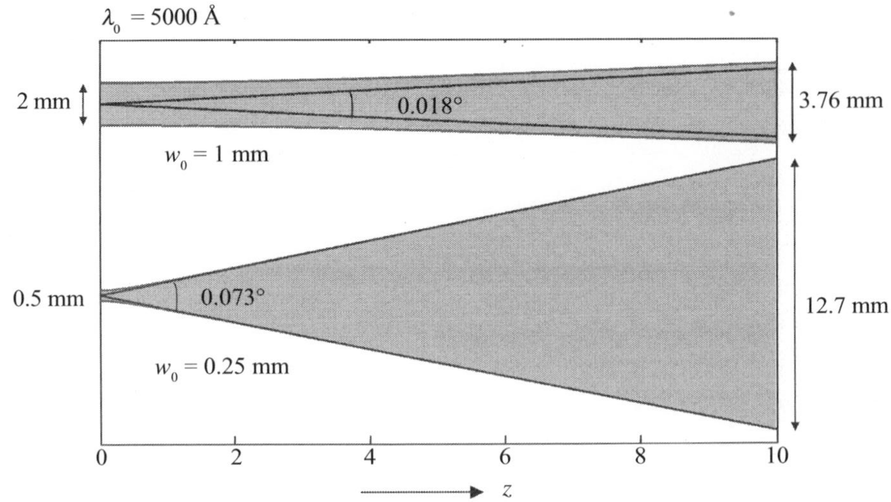

Fig. 2.8: Diffraction divergence of a Gaussian beam whose phase front is plane at $z = 0$. The figure shows the increase in the diffraction divergence as the initial spot size is decreased from 1 mm to 0.25 mm; the wavelength is assumed to be 5000 Å. Figure adapted from Ajoy Ghatak's *Optics*, 4th Edition, McGraw-Hill, New Delhi (2008).

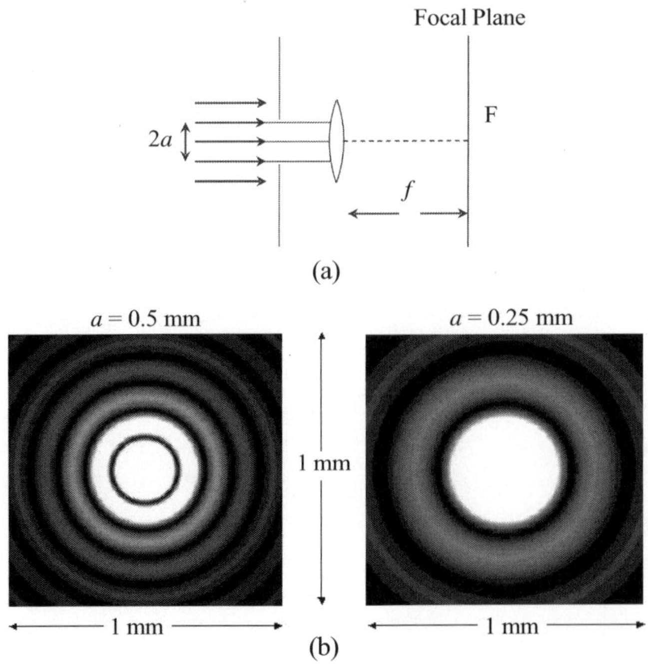

Fig. 2.9: (a) A truncated plane wave incident on a lens; (b) the resulting intensity distribution (which is known as the Airy pattern) at the focal plane of a lens. The computer generated Airy patterns correspond to $f = 20$ cm and $\lambda = 0.5\ \mu m$ and are adapted from Ajoy Ghatak's *Optics*, 4th Edition, McGraw-Hill, New Delhi (2008).

Thus, for, $a \approx 1$ cm, $\lambda \approx 6 \times 10^{-5}$ cm, $f \approx 10$ cm, 84% of light energy would be contained in the area

$$\pi\left(\frac{3.832\lambda f}{2\pi a}\right)^2 \approx 0.42 \times 10^{-6}\ \text{cm}^2$$

at the focal plane of the lens. Thus, a 1 W laser beam will produce an intensity of

$$\frac{1\ \text{W}}{0.42\times10^{-6}\,\text{cm}^2} \approx 2.4\ \text{MW/cm}^2$$

at the focal plane of the lens. The intensity distribution along the z-axis is given by

$$I(0, 0, z) = I_0\left[\frac{\sin(w/4)}{w/4}\right]^2 \tag{2.35}$$

where

$$w = \frac{2\pi a^2(\Delta z)}{\lambda f^2} \tag{2.36}$$

and Δz represents the axial distance from the focal point. At $w \approx \pm 3.2$

$$\left[\frac{\sin(w/4)}{w/4}\right]^2 \approx 0.8$$

implying that the intensity drops by about 20% for

$$\Delta z = z_0 \approx \pm\frac{3.2\lambda}{2\pi}\left(\frac{f}{a}\right)^2 \tag{2.37}$$

$$\approx 3 \times 10^{-3}\ \text{cm}$$

for $\lambda \approx 6 \times 10^{-5}$ cm, $f \approx 10$ cm, and $a \approx 1$ cm. The quantity z_0 is known as the *focal tolerance*.

2.8 The beam expander

In several experiments, often one uses a beam expander (see Fig. 2.10) to produce a beam of small focused spot size; however, the small focused spot would lead to a small depth of focus.

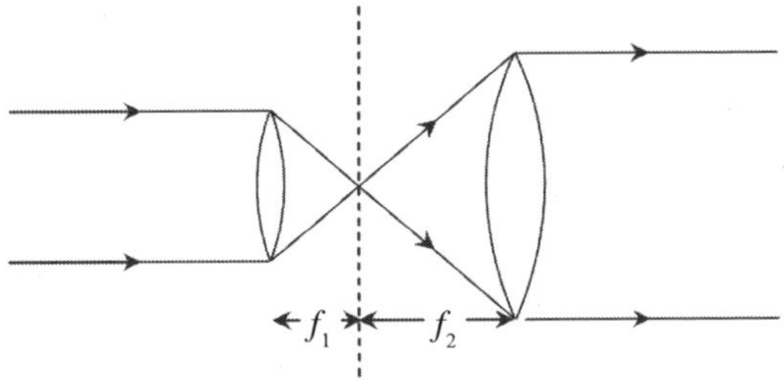

Fig. 2.10: Two convex lenses separated by a distance equal to the sum of their focal lengths act as a beam expander.

References

1. Ajoy Ghatak, *Optics*, 4th Edition, McGraw-Hill, New Delhi (2008).
2. Ajoy Ghatak and K. Thyagarajan, *Optical Electronics*, Cambridge University Press, Cambridge (1989). Reprinted by Foundation Books, New Delhi (1991).
3. O.S. Heavens and R.W. Ditchburn, *Insight into Optics*, John Wiley and Sons, Chichester (1991).
4. E. Hecht and A. Zajac, *Optics*, 2nd Edition, Addison-Wesley, Reading, MA (1987).
5. F.A. Jenkins and H.E. White, *Fundamentals of Optics*, 4th Edition, McGraw-Hill, New York (1976).
6. M. Born and E. Wolf, *Principles of Optics*, 6th Edition, Pergamon Press, New York (1980).

3

Basic Fiber Optics

Ajoy Ghatak

3.1 Introduction

An optical fiber is a cylindrical structure consisting of a glass core, which is surrounded by a cladding of slightly lower refractive index. The core diameter can vary from about 5 to 100 μm depending on applications and the cladding diameter is usually a standard 125 $\pm$ 1 μm. For preserving its pristine strength and robustness as well as greater protection from the changes in the environment of the fiber, there is usually a soft plastic primary coating whose diameter is about 250 μm, which is often followed by another layer of secondary coating ~1 mm diameter (see Fig. 3.1). Almost all present day communication grade fibers are essentially made of silica (SiO_2) as the basic raw material with few percent of dopants like germania, phosphorus, fluorine as refractive index modifiers.

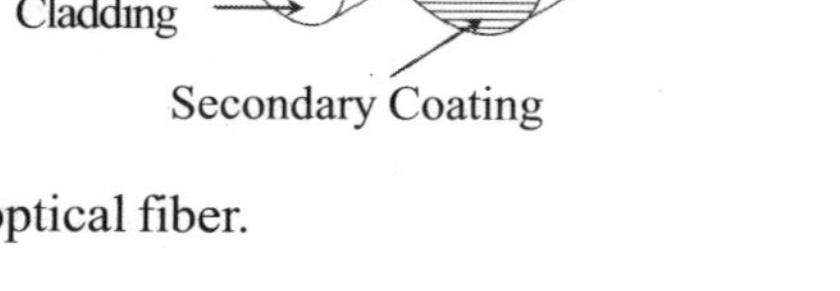

Fig. 3.1: Geometry of an optical fiber.

In certain specific applications such as sensing, plastic (like PMMA) fibers or plastic-clad silica fibers are also used. In this chapter we will briefly describe the ray analysis for light propagation in an optical fiber along with some elementary concepts in loss calculations. For further details on basic fiber optics the reader may refer to the list of books appended at the end of this chapter.

3.2 Snell's laws and total internal reflection

The refractive index of a medium is defined through the equation

$$n = \frac{c}{v} \tag{3.1}$$

where c is the velocity of light in vacuum, and v is the velocity of light in that medium. Let n_1 and n_2 represent the refractive indices of two semi-infinite media forming an interface. If $n_2 > n_1$, the second medium is said to be optically denser than the first medium. One could also say that the first medium is optically rarer than the second medium.

When a light ray propagates across an interface between two optically different media, a portion of the incident energy is reflected back into the first medium and a portion is refracted into the second medium. If the second medium is optically denser than the first medium (for example, when a ray goes from air into glass), the refracted ray is bent towards the normal to the interface. On the other hand, when the ray goes into an optically rarer medium (for example, from glass into air), it bends away from this normal [see Fig. 3.2(b)]. In either case, according to the Snell's law,

$$n_1 \sin \phi_1 = n_2 \sin \phi_2 \tag{3.2}$$

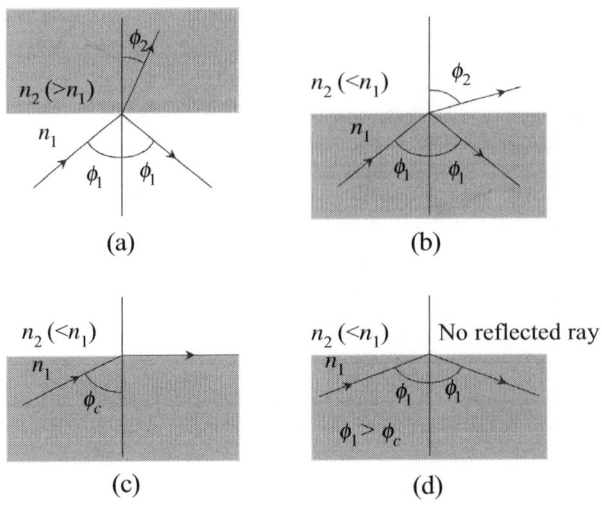

Fig. 3.2: Reflection and refraction of a plane wave at an interface between two semi-infinite dielectrics.

where ϕ_1 and ϕ_2 represent the angles of incidence and refraction, respectively. Referring to Fig. 3.2(b), we find that if the angle of incidence ϕ_1 is increased, the refracted ray bends more and more away from the normal, and at a particular angle of incidence – known as the *critical angle*, ϕ_c – the refracted ray becomes perpendicular to the normal, i.e. it *grazes* along the interface [see Fig. 3.2(c)]. When the angle of incidence is increased beyond this critical angle, no refracted ray exists and all the incident energy is reflected back into the optically denser medium [see Fig. 3.2(d)]. This phenomenon wherein all the incident energy is reflected back is called as *total internal reflection*. The critical angle ϕ_c is given by [cf. Eq. (3.2)]

$$\phi_c = \sin^{-1}\left(\frac{n_2}{n_1}\right) \tag{3.3}$$

3.3 The optical fiber

An optical fiber in its simplest form consists of a cylindrical central core region of refractive index n_1, cladded by a material of slightly lower refractive index, n_2 (see Fig. 3.3). The refractive index distribution (in the transverse direction) is thus given by

$$n(r) = \begin{cases} n_1 & 0 < r < a \quad \text{Core} \\ n_2 & r > a \quad \text{Cladding} \end{cases} \tag{3.4}$$

where a is the radius of the core. For a typical multimode telecom grade silica fiber[1]

$$n_1 \approx 1.48,\ n_2 \approx 1.46 \text{ and } a \approx 25\ \mu m \tag{3.5}$$

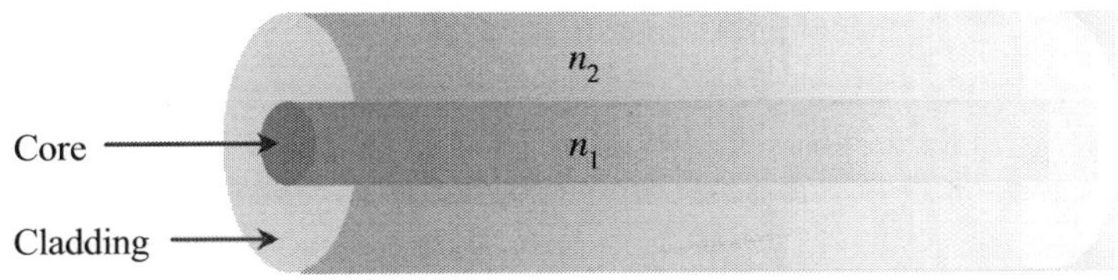

Fig. 3.3: An optical fiber, which consists of a cylindrical core, cladded by a material of a slightly lower refractive index ($n_2 < n_1$).

[1] Single-mode fibers will be discussed later.

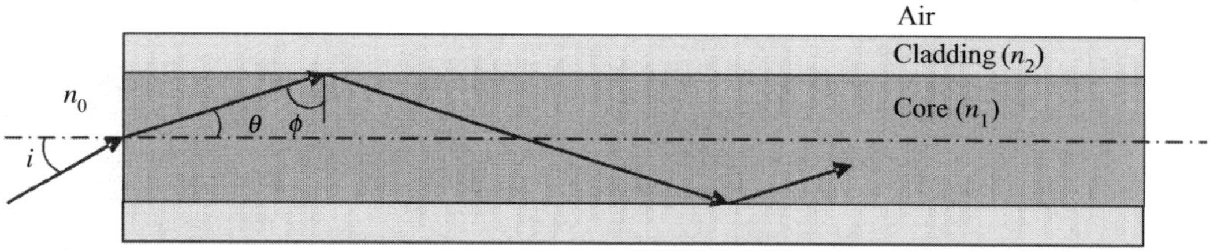

Fig. 3.4: Light guidance takes place through the phenomenon of total internal reflection at the core-cladding interface.

Fig. 3.5: A step-index multimode fiber illuminated by He-Ne laser with bright output light spot. The light coming out of the optical fiber is primarily due to Rayleigh scattering; a colour photo appears on the cover of the book. [The fiber was produced at the fiber drawing facility at CGCRI, Kolkata; photograph courtesy Dr Shyamal Bhadra and Ms Atasi Pal].

Figure 3.4 shows the longitudinal cross-section of a step-index multimode fiber. For a ray entering the fiber, if the angle of incidence ϕ (at the core-cladding interface) is greater than the critical angle, ϕ_c (= $\sin^{-1} n_2/n_1$) then the ray will undergo total internal reflection at that interface. Further, because of the cylindrical symmetry in the fiber structure, this ray will suffer total internal reflection at the lower interface also, and therefore get guided through the core by repeated total internal reflections.

Thus, an optical fiber acts as a "light guide" and it is therefore also known as an optical waveguide. Figure 3.5 shows the actual guidance of a light beam as it propagates through a long optical fiber.

3.4 The numerical aperture

Referring to Fig. 3.4, consider a ray, which is incident on the entrance aperture of the fiber making an angle i with the axis. Assuming the outside medium to have a refractive index n_0 (which for most practical cases is air, whose refractive index is unity), we get

$$\frac{\sin i}{\sin\theta} = \frac{n_1}{n_0} \tag{3.6}$$

Obviously, if this ray has to suffer total internal reflection at the core-cladding interface,

$$\sin\phi(=\cos\theta) > \frac{n_2}{n_1} \tag{3.7}$$

or

$$\sin\theta < \left[1-\left(\frac{n_2}{n_1}\right)^2\right]^{1/2} \tag{3.8}$$

$$\sin i < \frac{n_1}{n_0}\left[1-\left(\frac{n_2}{n_1}\right)^2\right]^{1/2} = \left[\left(\frac{(n_1^2-n_2^2)}{n_0^2}\right)\right]^{1/2} \tag{3.9}$$

Assuming $n_0 = 1$, the maximum value of sin i for a ray to be guided is given by

$$\sin i_m = \begin{cases}(n_1^2-n_2^2)^{1/2} & \text{if } n_1^2 < n_2^2+1\\ 1 & \text{if } n_1^2 > n_2^2+1\end{cases} \tag{3.10}$$

Thus, if a cone of light is incident on one end of the fiber, it will be guided through it provided the semi-angle of the cone is less than i_m. The quantity sin i_m is known as the numerical aperture (NA) of the fiber and is a measure of the *light gathering power* of the fiber.

For an optical fiber, in almost all practical situations, $n_1^2 < n_2^2 + 1$, and therefore one defines the numerical aperture of the fiber by the following equation:

$$NA = (n_1^2-n_2^2)^{1/2} \tag{3.11}$$

Thus, for $n_1 = 1.49$ and $n_2 = 1.46$, NA $\approx$ 0.30 implying that the acceptance angle is

$$i_m = \sin^{-1}(NA) = 17.3° \ (\approx 0.30 \text{ rad}) \tag{3.12}$$

In order to calculate the solid acceptance angle $\boldsymbol{\Omega}$ (see Fig. 3.6) we note that

$$\text{Area A} = \pi(r\sin i_m)^2 \tag{3.13}$$

Thus, for small i_m, solid acceptance angle $\Omega = \dfrac{A}{r^2} \approx \pi\, i_m^2 \approx 0.29$ steradians.

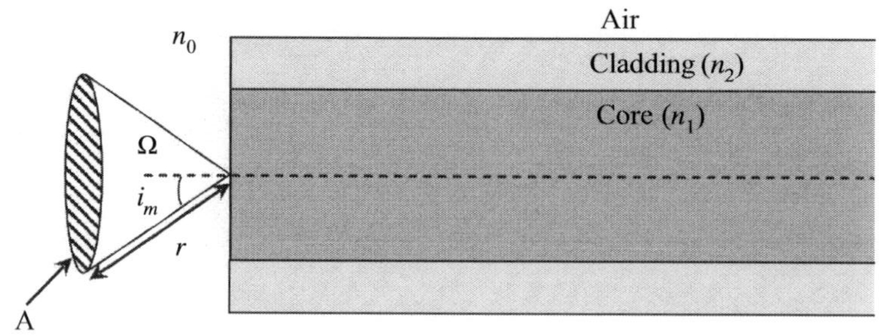

Fig. 3.6: Representation of the "light acceptance cone" of a solid angle Ω at the input end of the optical fiber.

3.5 Attenuation

If P_1 and P_2 represent the input and output powers, then

$$\text{Loss of Power in dB, } \alpha = 10 \log_{10} \frac{P_1}{P_2} \tag{3.14}$$

Thus, in the configuration shown in Fig. 3.7,

The input power, $P_1 = 100\ \mu$W

The output power, $P_2 = 15\ \mu$W

$\Rightarrow$ Total attenuation in traversing 5 km length of the fiber $= 10 \log\left(\dfrac{100}{15}\right) \approx 8.24$ dB

or, Loss in the fiber ≈ 1.65 dB/km

On the other hand, a loss of 0.2 dB/km would imply a loss of only 4.5% of power ($P_2/P_1 = 0.955$) in a distance of 1 km.

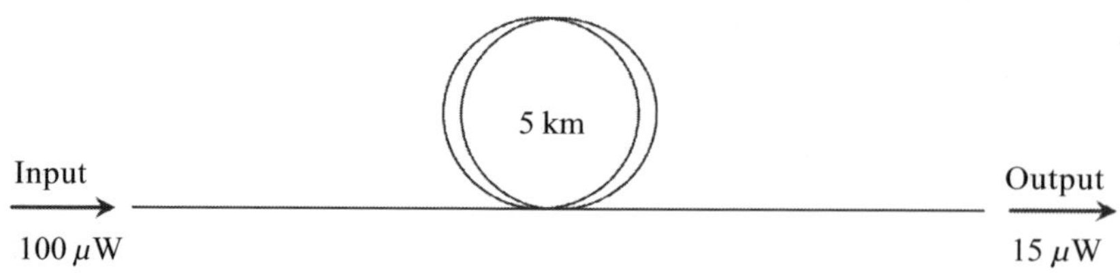

Fig. 3.7: A fiber loop of length 5 km carrying optical power.

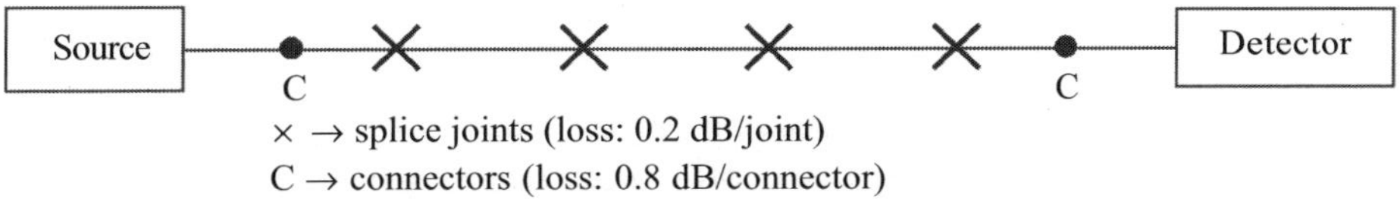

Fig. 3.8: Schematic of a fiber link having connectors and splice joints.

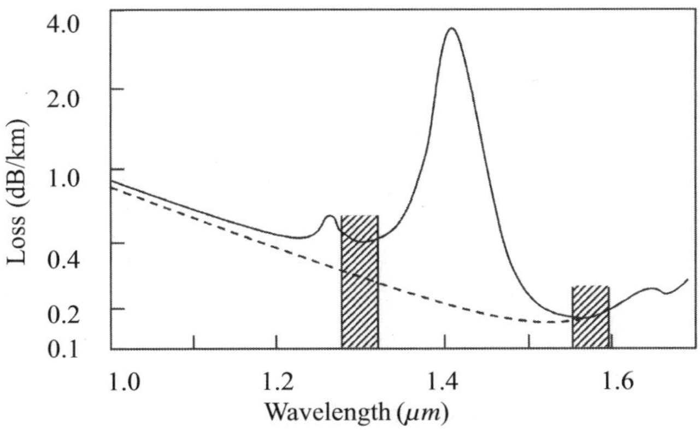

Fig. 3.9: Typical spectral response of a high silica optical fiber. The dotted line represents the intrinsic loss due to the Rayleigh scattering.

In order to illustrate the advantage of power loss calculations in dB, we consider a 50 km long fiber optic link with 2 connectors and 4 splice joints as shown in Fig. 3.8. If the fiber is characterized with a loss of 0.2 dB/km, the loss at each splice is 0.2 dB (i.e., 4.5% of power is lost at each splice) and the loss at a connector is 0.8 dB, then the total loss is given by

Loss at 4 splice joints $= 4 \times 0.2$ dB $= 0.8$ dB,

Loss at the 2 connectors $= 2 \times 0.8$ dB $= 1.6$ dB,

Loss in traversing 50 km of the optical fiber $= 50$ km $\times\ 0.2$ dB/km $= 10$ dB.

Hence,

$$\text{Total loss} = (0.8 + 1.6 + 10)\ \text{dB} = 12.4\ \text{dB}.$$

i.e.

$$10 \log \frac{P_1}{P_2} = 12.4\ \text{dB}$$

or,

$$= 17.38$$

Therefore, for a 100 μW source power coupled into the fiber (P_1) the power received at the detector ($=P_2$) would be $\approx 5.75\ \mu$W.

The attenuation curve as a function of wavelength for a typical silica fiber is shown in Fig. 3.9. The peaks at $\lambda \approx 0.8\ \mu m$ and $1.39\ \mu m$ are due to the presence of traces of metal impurities and OH^- ions, respectively. One can see that the loss is very small (≈0.5 dB/km) at $\lambda = 1.30\ \mu m$, and is still smaller (≈0.2 dB/km) at $\lambda = 1.55\ \mu m$. It so happens that in optical fibers, the dispersion is also very small (close to zero) at $1.3\ \mu m$ at which loss is ≈0.34 dB/km. Therefore many of the present day communication systems operate at $\lambda \approx 1.3\ \mu m$; this wavelength region (shown as a vertical bar in Fig. 3.9) is referred to as the "$1.3\ \mu m$ window". Obviously, it would be better if we could operate at $\lambda \approx 1.55\ \mu m$, where losses are even smaller (by about 40%). The dispersion can also be made very small at $\lambda \approx 1.55\ \mu m$ by reducing the core radius and increasing the core refractive index or through complex refractive index profiles, e.g. a segmented core profile. These fibers are referred as *dispersion shifted fibers*. In addition, there is another class of fibers known as non-zero dispersion shifted fibers (NZDSF), which are purposely designed to exhibit a small but finite dispersion at $\lambda \approx 1.55\ \mu m$ in order to contain certain nonlinear optical propagation effects (which are beyond the scope for discussion in this book) (see e.g. Ref. 7). Due to the development of what are known as Erbium doped fiber amplifiers (EDFA), which can amplify signals over a broad range of λ (1530 ~ 1565 nm), indeed many of the new optical fiber communication systems now operate at the "$1.55\ \mu m$ window".

3.6 The dBm scale

It is often very convenient to measure the power level of a beam in dBm which is defined as

$$P(\text{dBm}) = 10 \log_{10} P(\text{mW}) \tag{3.15}$$

Thus,

$$1\ \text{mW} \Leftrightarrow 0\ \text{dBm}$$

$$1\ \text{W} \Leftrightarrow 30\ \text{dBm}$$

$$1\ \text{mW} \Leftrightarrow -30\ \text{dBm}$$

$$1\ \text{nW} \Leftrightarrow -60\ \text{dBm}$$

and,

$$0.4\ \text{W} = 400\ \text{mW} \Leftrightarrow \approx 26\ \text{dBm}$$

Using the dBm scale Eq. (3.14) becomes

$$\text{Loss in dB, } \alpha = P_{input}\ (\text{dBm}) - P_{output}\ (\text{dBm}) \tag{3.16}$$

or,

$$P_{output}\ (\text{dBm}) = P_{input}\ (\text{dBm}) - \alpha\ (\text{dB}) \tag{3.17}$$

Because of the above equation, calculation of the drop in power levels using the dBm scale become very easy. As an example, we consider a 10 mW laser beam passing through a 50 km fiber link of loss 0.25 dB/km. There are 5 connectors in the link – each having a loss of 1 dB; thus the total loss in the link is $50 \times 0.25 + 5 = 17.5$ dB. Since the input power is 10 dBm, the power at the output would be –7.5 dBm which is equal to 0.178 mW.

References

1. Ajoy Ghatak and K. Thyagarajan, *Introduction to Fiber Optics*, Cambridge University Press, Cambridge (1998). Reprinted by Foundation Books, New Delhi.
2. K. Thyagarajan and Ajoy Ghatak, *Fiber Optic Essentials*, John Wiley and Sons, New York (2007).
3. P.K. Cheo, *Fiber Optics and Optoelectronics*, 2nd Edition, Prentice-Hall International, New Jersey (1990).
4. A.C. Cherin, *An Introduction to Optical Fibers*, McGraw-Hill International, Japan (1983).
5. Y. Suematsu and K. Iga, *Introduction to Optical Fiber Communications*, John Wiley and Sons, New York (1982).
6. M.J. Adams, *An Introduction to Optical Waveguides*, John Wiley and Sons, New York (1981).
7. B.P. Pal (Ed.), *Guided Wave Optical Components and Devices*: *Basics, Technology, and Applications*, Academic Press/Elsevier, Burlington, 2006.

4

Modes of an Optical Waveguide

Ajoy Ghatak

4.1 Introduction

In the last chapter we presented a simple picture of light propagation in optical fibers using the concept of total internal reflection. As we know, light is an electromagnetic wave, and therefore a more rigorous study of the propagation characteristics of optical fibers requires knowledge of the propagating fields, which are known as "modes" of the fiber. An optical waveguide, by definition, is an optical structure, which is capable of guiding an optical wave along its length. In general, optical waveguides are made up of dielectric materials having very low absorption at optical wavelengths. In this chapter we introduce the concept of modes in an optical waveguide, and describe the propagation of light in terms of these modes in optical fibers.

4.2 Modes of a planar waveguide

A general dielectric waveguide is described by the refractive index variation

$$n^2 = n^2 (x, y) \tag{4.1}$$

Because of n^2 depending on x and y only, the solution of Maxwell's equations can be described by the following equation (see e.g. References 1–3):

$$E(x, y, z, t) = E_0 (x, y) e^{i(\omega t - \beta z)} \tag{4.2}$$

The above equation defines the modes of a waveguide in which the transverse field distributions remain unchanged (except for a change in phase) as they propagate along the waveguide. The simplest optical waveguide is a symmetric planar waveguide consisting of a dielectric film of refractive index n_1 (of thickness ≈ few microns), sandwiched between two semi-infinite layers of another dielectric of slightly lower refractive index n_2 ($<n_1$) as shown in Fig. 4.1. In fact, planar waveguides or thin film waveguides are the building blocks of an important related field known as *Integrated Optics*. The concept of "modes" of an optical waveguide can be best understood by considering a step-index planar waveguide. Mathematically, modes of a waveguide are the allowed solutions to the boundary value problem of propagation in the waveguide. Since light is an electromagnetic wave, these solutions are nothing but the various possible electric and magnetic field distributions.

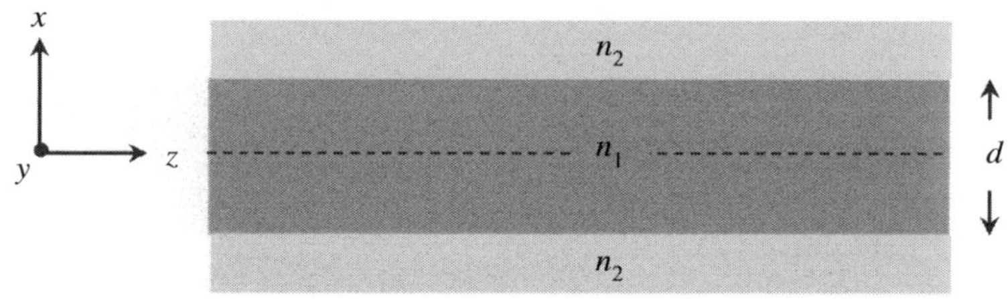

Fig. 4.1: A planar dielectric waveguide of thickness d (along x direction) but infinitely extended along the y direction. Light propagates along the z direction.

It can be shown that a planar optical waveguide supports two distinct sets of modes known as transverse electric (TE-) and transverse magnetic (TM-) modes (see e.g. Chapter 7 of Ref. 2). As the names indicate, for TE modes the electric field vector is always transverse (i.e. perpendicular) to the direction of propagation, and for TM modes the magnetic field vector is always transverse to the direction of propagation. Further, referring to Fig. 4.1, since the refractive index distribution is symmetric about the waveguide axis, the waveguide supports what are known as 'even' (symmetric) and 'odd' (anti-symmetric) modes. To appreciate the nature of various mode field distributions, we consider the complete field solutions for the TE modes of a symmetric slab waveguide:

Even modes
$$\psi(x,z,t) = \begin{cases} A\cos \kappa x \, e^{i(\omega t-\beta z)} & |x| < \dfrac{d}{2} \\ A\cos\left(\dfrac{\kappa d}{2}\right) e^{-\gamma\left(|x|-\frac{d}{2}\right)} e^{i(\omega t-\beta z)} & |x| > \dfrac{d}{2} \end{cases} \tag{4.3}$$

Odd modes
$$\psi(x,z,t) = \begin{cases} B\sin \kappa x \, e^{i(\omega t-\beta z)} & |x| < \dfrac{d}{2} \\ B\sin\left(\dfrac{\kappa d}{2}\right) e^{-\gamma\left(|x|-\frac{d}{2}\right)} e^{i(\omega t-\beta z)} & |x| > \dfrac{d}{2} \end{cases} \tag{4.4}$$

where
$$\kappa = (k_0^2 n_1^2 - \beta^2)^{1/2}, \quad \gamma = (\beta^2 - k_0^2 n_2^2)^{1/2} \tag{4.5}$$

β represents the propagation constants of the symmetric and anti-symmetric modes, and are discrete solutions of the following transcendental equations:

$$\tan\left(\frac{\kappa d}{2}\right) = \frac{\gamma}{\kappa} \qquad \text{symmetric modes}$$

$$\tan\left(\frac{\kappa d}{2}\right) = -\frac{\kappa}{\gamma} \qquad \text{anti-symmetric modes} \tag{4.6}$$

We may mention here that for a waveguiding structure shown in Fig. 4.1, guided modes will correspond to

$$n_2 < \frac{\beta}{k_0} < n_1 \tag{4.7}$$

which will lead to an "oscillatory" solution in the core and an exponentially decaying solution in the cladding. Note that the fields have no y-dependence since the waveguide is infinitely extended in the y direction. Physically, it implies that the field is free to diffract along this direction. The oscillatory term $\exp[i(\omega t - \beta z)]$ represents the phase variation as the wave propagates along the z-axis. Thus, for example, the transverse field distribution of the TE even modes are given by

$$\psi(x) = \begin{cases} A\cos\kappa x & |x| < \frac{d}{2} \\ A\cos\left(\frac{\kappa d}{2}\right) e^{-\gamma\left(|x| - \frac{d}{2}\right)} & |x| > \frac{d}{2} \end{cases} \tag{4.8}$$

which is independent of the propagation distance z. Figure 4.2 shows the field distributions corresponding to the first two even and odd TE modes of a waveguide with the following parameters: $n_1 = 1.49$, $n_2 = 1.46$, $d = 4\ \mu m$, $\lambda = 0.6328\ \mu m$. The corresponding propagation constants are

$$\frac{\beta}{k_0} = \begin{cases} 1.4885 & \text{For TE}_0 \text{ mode (even)} \\ 1.4839 & \text{For TE}_1 \text{ mode (odd)} \\ 1.4765 & \text{For TE}_2 \text{ mode (even)} \\ 1.4668 & \text{For TE}_3 \text{ mode (odd)} \end{cases} \tag{4.9}$$

In order to have a physical understanding of the modes, we consider the electric field pattern inside the film $\left(-\frac{d}{2} < x < \frac{d}{2}\right)$. Thus, for a symmetric TE mode

$$\psi(x, z, t) = A \cos \kappa x \, e^{i(\omega t - \beta z)}$$

$$= \frac{1}{2} A e^{i[\omega t - (\kappa x + \beta z)]} + \frac{1}{2} A e^{i[\omega t - (-\kappa x + \beta z)]} \tag{4.10}$$

Now, $$A\, e^{i(\omega t - [k_x x + k_y y + k_z z])} = A\, e^{i(\omega t - \mathbf{k.r})} \tag{4.11}$$

represents a plane wave propagating along the direction of $\boldsymbol{k}$, whose x, y and z components are k_x, k_y and k_z, respectively. Thus, for the two terms on the RHS of Eq. (4.10), we have

$$k_x = \kappa,\ k_y = 0,\ k_z = \beta \ \text{ and } \ k_x = -\kappa,\ k_y = 0,\ k_z = \beta \tag{4.12}$$

The first term on the RHS of Eq. (4.10), therefore, represents a plane wave with the propagation vector in the x-z plane making an angle θ with the z-axis (see Fig. 4.3), where

$$\tan\theta = \frac{k_x}{k_z} = \frac{\kappa}{\beta} \tag{4.13}$$

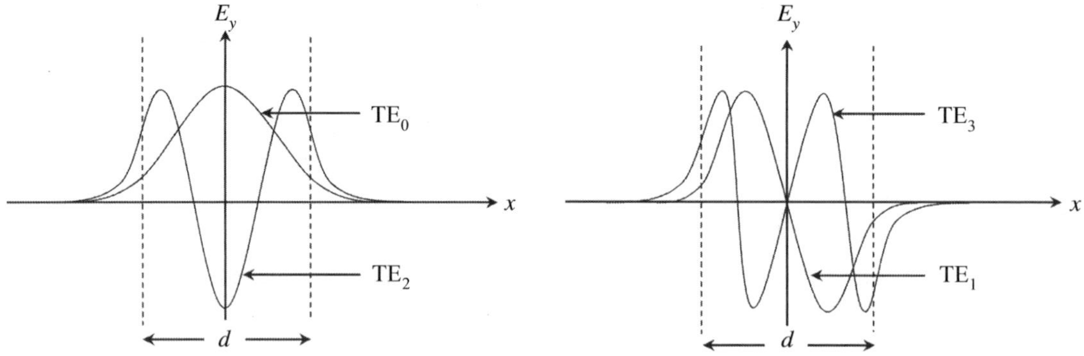

Fig. 4.2: TE mode field distributions in a step-index planar waveguide with $n_1 = 1.49$, $n_2 = 1.56$, $a = 4\ \mu m$ and $\lambda = 0.6328\ \mu m$. TE_0 and TE_2 are known as *even* modes, while TE_1 and TE_3 are the *odd* modes.

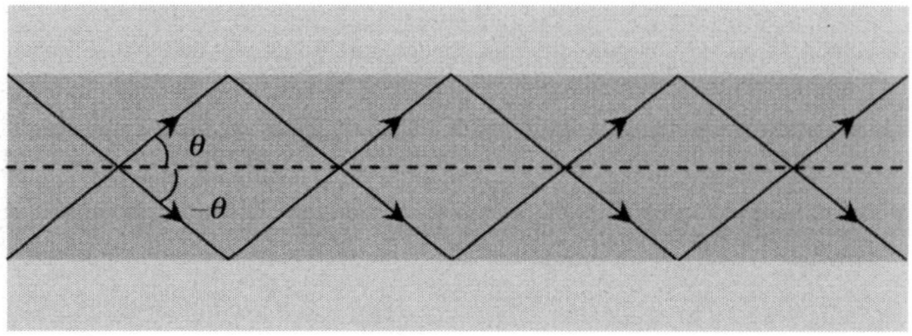

Fig. 4.3: A guided mode in a step-index waveguide corresponds to the superposition of two plane waves propagating at particular angles $\pm\theta$ with the z-axis.

Similarly, the second term represents a plane wave making an angle $-\theta$ with the z-axis. Thus, we see that a guided mode can be considered to be a superposition of a pair of plane waves, which propagate at angles $\pm\theta$ with the z-axis (i.e. propagating upward and downward). Since only discrete values of β are allowed (which we designate as β_m), only discrete angles of propagation of plane waves (or of the rays) are allowed. That is, each mode is characterized by a discrete angle of propagation θ_m of the rays inside the waveguide. Using the relation $\sec^2\theta$ $(=1+\tan^2\theta)$, Eq. (4.13) can be written in the form

$$\cos\theta = \frac{\beta}{(\beta^2+\kappa^2)^{1/2}} = \frac{\beta}{k_0 n_1} \tag{4.14}$$

where we have used Eq. (4.5). Thus

$$\theta_m = \cos^{-1}\left(\frac{\beta_m}{k_0 n_1}\right) \tag{4.15}$$

Corresponding to Eq. (4.9)

$$\theta_1 \approx 2.6^0;\ \theta_2 \approx 5.2^0;\ \theta \approx 7.7^0;\ \theta_4 \approx 10.1^0 \tag{4.16}$$

Note that for guided modes β_m cannot have a value less than $k_0 n_2$. For $\beta = k_0 n_2$ implies

$$\sin\phi = \sin(90-\theta) = \cos\theta = \frac{n_2}{n_1} \tag{4.17}$$

which defines the critical angle ϕ_c for total internal reflection at the core-cladding interface (see Sec. 3.2). $\beta = k_0 n_2$ is also known as the mode *cutoff condition*, i.e. the wave ceases to be a guided mode below its own cutoff.

From the above discussion it is clear that if we excite plane waves traveling at a particular mode angle θ_m, corresponding to a mode propagation constant β_m, then only that particular mode will be excited, even though the waveguide can support many other modes. Indeed, it is possible to selectively excite a mode or a group of modes of the waveguide. A technique for selective excitation of guided modes of an optical fiber is described in Chapter 18.

4.3 LP modes of a fiber

The modes of an optical fiber characterized by a cylindrically symmetric refractive index profile

$$n = n(r) \tag{4.18}$$

are given by (see e.g. Ref. 2)

$$\psi(r,\ \phi,\ z,\ t) = R_{l,\,m}(r)\begin{Bmatrix}\cos l\phi\\ \sin l\phi\end{Bmatrix}\exp[i(\omega t-\beta_{lm}z)] \tag{4.19}$$

where $l = 0, 1, 2, \ldots; m = 1, 2, 3, \ldots$ and β_{lm} represents the discrete propagation constant of the $(lm)^{th}$ mode. According to the above equation, for a particular value of β, there is a specific transverse field distribution which remains unchanged with propagation along the z-axis (which is the axis of the waveguide) except for a change in phase (because of the exponential term); these field distributions are referred to as the modes of the optical fiber. The functions $R_{lm}(r)$ are solutions of the following equation:

$$\frac{d^2R}{dr^2} + \frac{1}{r}\frac{dR}{dr} + \left[k_0^2 n^2(r) - \beta^2 - \frac{l^2}{r^2}\right]R(r) = 0 \tag{4.20}$$

where
$$k_0 = \frac{\omega}{c} \tag{4.21}$$

represents the free space wave number.

4.3.1 Step-index fiber

For a step-index fiber

$$n(r) = \begin{cases} n_1 & 0 < r < a \\ n_2 & r > a \end{cases} \tag{4.22}$$

where a represents the core radius. The solution of Eq. (4.20) would be given by (see e.g. Ref. 1)

$$R(r) = \begin{cases} \dfrac{A}{J_l(U)} J_l\left(U\dfrac{r}{a}\right), & 0 < r < a \\ \dfrac{A}{K_l(W)} K_l\left(W\dfrac{r}{a}\right), & r > a \end{cases} \tag{4.23}$$

where J_l and K_l represent the Bessel functions and the modified Bessel functions respectively, and

$$U = a\sqrt{k_0^2 n_1^2 - \beta^2} = V\sqrt{1-b}$$

$$W = a\sqrt{\beta^2 - k_0^2 n_2^2} = V\sqrt{b}$$

$$V = \sqrt{U^2 + W^2} = k_0 a\sqrt{n_1^2 - n_2^2} \tag{4.24}$$

$$b = \frac{\dfrac{\beta^2}{k_0^2} - n_2^2}{n_1^2 - n_2^2} = \frac{W^2}{V^2}$$

The parameters V and b are usually referred to as the "normalized waveguide parameters" (or the V-number) and "normalized propagation constant", respectively. Continuity of $R(r)$ and its derivative at $r = a$ (i.e. at the core-cladding interface) leads to the following transcendental equation:

$$V\sqrt{1-b}\,\frac{J_{l-1}(V\sqrt{1-b})}{J_l(V\sqrt{1-b})} = -V\sqrt{b}\,\frac{K_{l-1}(V\sqrt{b})}{K_l(V\sqrt{b})};\quad l \geq 1 \tag{4.25}$$

and
$$V\sqrt{1-b}\,\frac{J_1(V\sqrt{1-b})}{J_0(V\sqrt{1-b})} = V\sqrt{b}\,\frac{K_1(V\sqrt{b})}{K_0(V\sqrt{b})};\quad l = 0 \tag{4.26}$$

For a given value of V, the roots of the above equations will give us the values of the discrete propagation constants b_{lm} (and hence β_{lm}). In general, for a given value of l, there can be more than one root, which is designated by the subscript "m". The above results are valid when

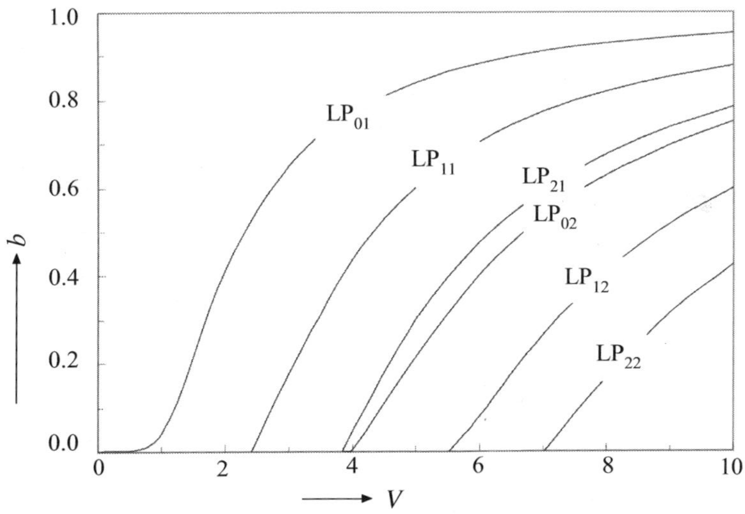

Fig. 4.4: Variation of the normalized propagation constant b with normalized waveguide parameter V corresponding to a few lower order modes.

$n_1 \approx n_2$, which is usually referred to as the "weakly guiding approximation". In this approximation the modes are designated as LP_{lm} modes; LP standing for "linearly polarized". For each mode, we can have two polarization states (x-polarized and y-polarized) corresponding to the same value of b_{lm} (or β_{lm}).

The variation of the normalized propagation constant b with V for a step-index fiber is shown in Fig. 4.4. As can be seen for $0 < V < 2.4048$, only the LP_{01} mode will propagate in the fiber as a guided mode. Similarly,

for $2.4048 < V < 3.8317$, only LP_{01} and LP_{11} modes will exist;
for $3.8317 < V < 5.1356$, only LP_{01}, LP_{11}, LP_{21} and LP_{02} modes will exist;
for $5.1356 < V < 5.5201$, only LP_{01}, LP_{11}, LP_{21}, LP_{02} and LP_{31} modes will exist.

The numbers limiting the V-value correspond to the zeros of the Bessel functions of various orders. The curves in Fig. 4.4 are universal, i.e. for a given step-index fiber and a given operating wavelength, we have to first calculate the value of V and then "read off" the corresponding values of b from the curves; the values of β can then be calculated by using the following relation:

$$\beta_{lm} = k_0\left[n_2^2 + b_{lm}^2(n_1^2 - n_2^2)\right]^{\frac{1}{2}} \tag{4.27}$$

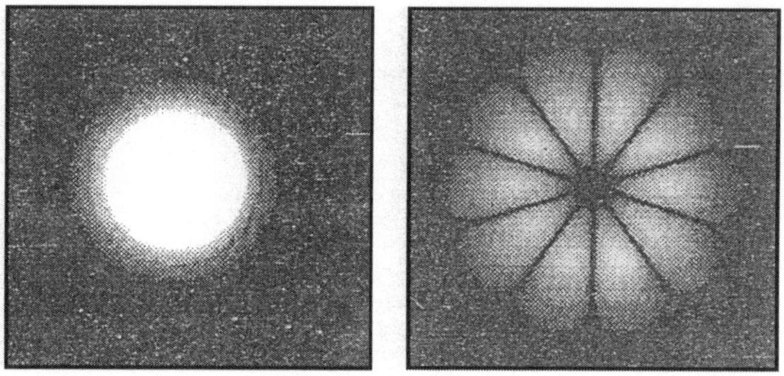

Fig. 4.5: Plots of computer generated intensity distribution of LP_{01} and LP_{51} modes of a particular step-index fiber. (Adapted from Ref. 5).

Since for guided modes

$$n_2 < \frac{\beta_{lm}}{k_0} < n_1, \tag{4.28}$$

we must have

$$0 < b_{lm} < 1 \tag{4.29}$$

Figure 4.5 shows the intensity distributions of the LP_{01} and LP_{51} modes; the values of l and m are obtained by using the following rules:

number of intensity zeros in the ϕ direction $= 2l$
number of intensity zeros in the radial direction $= m - 1$
(excluding the *zero* at $r = 0$)

As is obvious from the above analysis, and also from Fig. 4.4, for a step-index fiber with

$$0 < V < 2.4048 \tag{4.30}$$

we will have only one guided mode, namely the LP_{01} mode. Such a fiber is referred to as a "single-mode fiber" and is of tremendous interest in optical communication system.

It may be worthwhile to mention that in the region $1.5 \le V \le 2.5$, the value of b for the LP_{01} mode is quite accurately given by the following empirical formula [6]:

$$b \cong \left[1.1428 - \frac{0.996}{V}\right]^2 ; \quad 1.5 \le V \le 2.5 \qquad (4.31)$$

In Table 4.1 we have compared the values of b obtained by using the above formula with the numerical values obtained by using the software in Ref. 7. Furthermore, although the exact field is in terms of Bessel functions, a Gaussian distribution can often quite accurately represent the fundamental mode field

$$\psi(r) = A \exp\left[-\frac{r^2}{w^2}\right] \qquad (4.32)$$

with

$$\frac{w}{a} \cong 0.65 + \frac{1.619}{V^{3/2}} + \frac{2.879}{V^6}; \quad 0.8 \le V \le 2.5 \qquad (4.33)$$

where a represents the core radius and A gives the field value at $r = 0$. The above empirical relation is due to Marcuse [8], and has extensively been used for splice loss calculations, etc.

Table 4.1: Values of b as a function of V for a step-index fiber*.

V	b	b [using Eq. 4.31]
1.5	0.229248	0.229249
1.6	0.270063	0.270712
1.7	0.309467	0.310157
1.8	0.347068	0.347471
1.9	0.382660	0.382653
2.0	0.416163	0.415767
2.1	0.447581	0.446911
2.2	0.476969	0.476200
2.3	0.504416	0.503754
2.4	0.530026	0.529693
2.5	0.553915	0.554131

* The value of b (in the second column) is calculated by using the software in Ref. 7.

4.3.2 Graded index fiber

We next consider a highly multimoded parabolic index fiber characterized by the following refractive index variation (see Fig. 4.6)

$$n^2(r) = \begin{cases} n_1^2\left[1-2\Delta\left(\dfrac{r}{a}\right)^2\right]; & 0 < r < a \\ n_1^2\left[1-2\Delta\right] = n_2^2; & r > a \end{cases} \tag{4.34}$$

where

$$\Delta = \frac{n_1^2 - n_2^2}{2n_1^2} \tag{4.35}$$

For a highly multimoded fiber, $a \approx 25\text{–}50\ \mu m$ and $\Delta \approx 0.01$. For such a fiber, the propagation constants of the LP_{lm} modes are approximately given by

$$\beta_{lm}^2 \approx k_0^2 n_1^2 - 2(2m + l - 1)\frac{k_0 n_1\sqrt{2\Delta}}{a} \tag{4.36}$$

with $l = 0, 1, 2, \ldots$; $m = 1, 2, 3, \ldots\ldots$. Since, for guided modes

$$k_0^2 n_2^2 < \beta_{lm}^2 < k_0^2 n_1^2 \tag{4.37}$$

we must have

$$(2m + l - 1) < \frac{k_0\, a\, n_1\sqrt{2\Delta}}{2} = \frac{V}{2} \tag{4.38}$$

where once again

$$V = k_0 a\sqrt{n_1^2 - n_2^2} = k_0 a n_1\sqrt{2\Delta}$$

From Eq. (4.38) one can easily show that the total number of modes (including the 2 independent states of polarization) for a parabolic index fiber is given by

$$N \approx \frac{1}{4}V^2$$

The corresponding field patterns are given by

$$\psi_{lm}(r, \phi, z, t) = N_{lm} r^l \exp\left(-\frac{1}{2}\alpha^2 r^2\right) L_{m-1}^l(\alpha^2 r^2)\begin{Bmatrix}\cos l\phi \\ \sin l\varphi\end{Bmatrix} e^{i(\omega t - \beta_{lm} z)} \tag{4.39}$$

where

$$\alpha^2 = \frac{k_0 n_1}{a}\sqrt{2\Delta} = \frac{V}{a^2} \tag{4.40}$$

and L_{m-1}^l represents associated Laguerre polynomials; N_{lm} are the corresponding

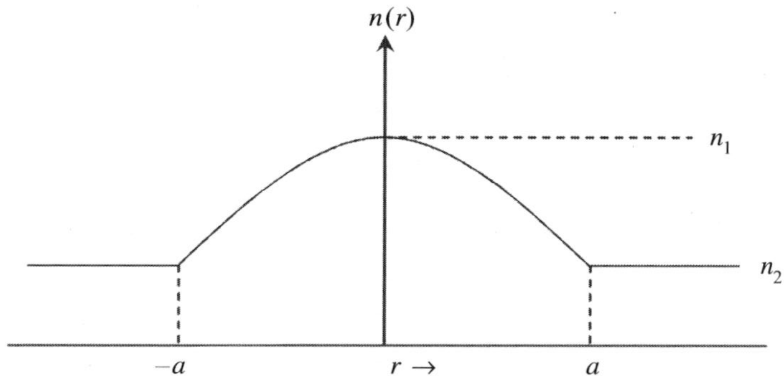

Fig. 4.6: Parabolic refractive index variation represented by Eq. (4.34).

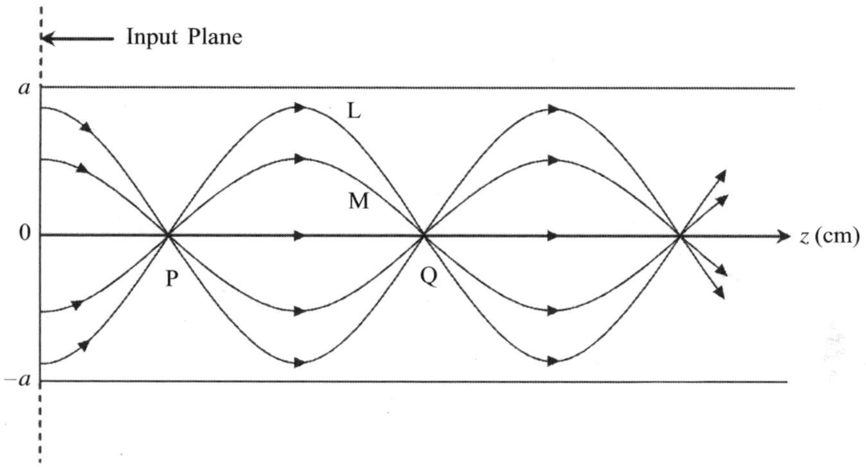

Fig. 4.7: Paraxial ray paths of meridional rays in a parabolic index waveguide.

normalization constants. The fundamental mode in this case corresponds to $l = 0$, $m = 1$, and has a Gaussian distribution

$$\psi_{01} = \exp\left[-\frac{1}{2}V\left(\frac{r}{a}\right)^2\right] \tag{4.41}$$

In the next section we shall describe the ray paths in a graded index multimode fiber, which again helps in the physical understanding of propagation in a graded index medium.

4.4 Ray paths in parabolic index fibers

The multimode graded index fibers are usually characterized by a parabolic refractive index distribution given by Eq. (4.34). The ray paths in such a medium are given by [see e.g. Ref. 2, 3, 8]

$$x(z) = A \sin \gamma z + B \cos \gamma z \tag{4.42}$$

and

$$y(z) = C \sin \gamma z + D \cos \gamma z \tag{4.43}$$

where

$$\gamma = \frac{n_1 \sqrt{2\Delta}}{a\tilde{\beta}} \tag{4.44}$$

and

$$\tilde{\beta} = n(r) \cos \theta(r) \tag{4.45}$$

is an invariant of the ray path, implying that as the ray propagates through the waveguide it would bend in such a way that the product $n(r) \cos \theta(r)$ would remain unchanged; θ is the angle that the ray makes with the z-axis. The constants A, B, C and D are determined from the initial launching conditions. For example, for the meridional rays shown in Fig. 4.7,

$$x(z) = A \sin \gamma z \tag{4.46}$$

with

$$A = \frac{a}{\sqrt{2\Delta}} \sin\theta_1 \tag{4.47}$$

and

$$\tilde{\beta} = n_1 \cos \theta_1 \tag{4.48}$$

where θ_1 is the angle that the ray makes with the axis. Different rays in Fig. 4.7 correspond to different values of θ_1 (and hence of the invariant $\tilde{\beta}$); however, their periodic length

$$z_p = \frac{2\pi}{\gamma} = \frac{2\pi\tilde{\beta}a}{n_1\sqrt{2\Delta}} = \frac{2\pi a}{\sqrt{2\Delta}} \cos\theta_1 \tag{4.49}$$

has a very weak dependence on θ_1 for *paraxial rays* (i.e. for small values of θ_1). Thus, rays launched at different angles with the axis ($\tilde{\beta} > n_2$) get trapped in the medium and hence the fiber acts as a waveguide. Further, in the paraxial approximation, since all rays emanating from P will focus at Q, and as a consequence of Fermat's principle, all rays will take the same amount of time. Physically, although the ray PLQ traverses a larger path in comparison to PMQ, it does so in a medium of "lower" average refractive index. Thus, the greater path length is compensated for by a greater "average speed" and hence *all* rays take approximately the same amount of time to propagate through a certain distance of the waveguide. This leads to extremely small pulse dispersions when such fibers are used for pulse transmission. Although present generation optical communication systems essentially are based on single-mode fibers, during mid-1970s graded index fibers were also widely used in such systems.

On the other hand, if the ray is launched in such a way that $B = C = 0$ and $A = D$, which implies that

$$x(z) = A \sin \gamma z, \qquad y(z) = A \cos \gamma z \tag{4.50}$$

$$\Rightarrow \qquad x^2(z) + y^2(z) = A^2 \tag{4.51}$$

which is independent of z. Thus, the ray spirals around the z-axis as a helix at a constant distance A from the z-axis (see Fig. 4.8). This is an extreme form of a skew ray!

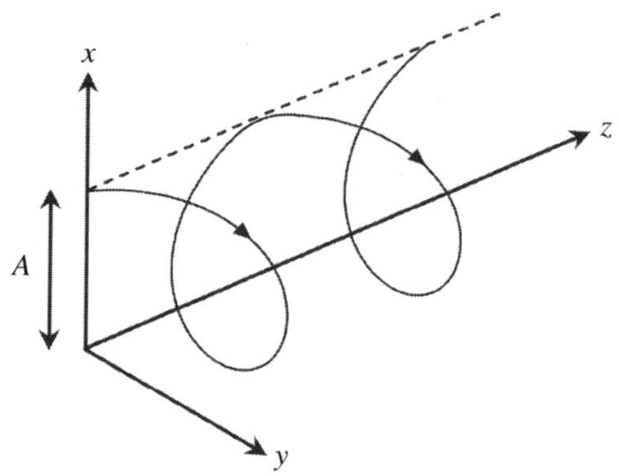

Fig. 4.8: A helical ray path in a parabolic index fiber.

4.5 Dispersion in optical fibers

In digital communication systems, information to be sent is first coded in the form of pulses and then these pulses of light are transmitted from the transmitter to the receiver where the information is decoded. Larger the number of pulses that can be sent per unit time and still be resolvable at the receiver end, larger would be the transmission capacity of the system. A pulse of light sent into a fiber broadens in time as it propagates through the fiber; this phenomenon is known as pulse dispersion and occurs primarily because of the following mechanisms:

1. In multimode fibers, different rays take different times to propagate through a given length of the fiber; we will discuss this for a step-index fiber and for a parabolic index fiber in this and the following sections. In the language of wave optics, this is known as *intermodal dispersion* because it arises due to different modes traveling with different velocities.
2. Any given light source emits over a range of wavelengths and, because of the intrinsic property of the material of the fiber, different wavelengths take different amounts of time to propagate along the same path. This is known as material dispersion and obviously, it is present in both single-mode and multimode fibers.

3. On the other hand, in single-mode fibers since there is only one mode, there is no intermodal dispersion; however, we have what is known as waveguide dispersion, which is due to the geometry of the fiber. Waveguide dispersion is present in multimode fibers also but the effect is very small and can be neglected.

4.5.1 Intermodal dispersion

Intermodal dispersion is due to the fact that different modes take different time to propagate through a certain length of the fiber. For multimode fibers, the ray analysis is applicable and one can show that the temporal broadening is given by the following expressions (see e.g. Ref. 2, 3 and 9):

$$\Delta\tau \approx \frac{n_1\Delta}{c}L \quad \text{for a step-index fiber,} \tag{4.52}$$

and

$$\Delta\tau \approx \frac{n_2 L}{2c}\Delta^2 \quad \text{for a parabolic index fiber.} \tag{4.53}$$

In the above equations, c is the speed of light in free space ($\approx 3 \times 10^8$ m/s), L is the length of the fiber, and

$$\Delta = \frac{n_1 - n_2}{n_2} \tag{4.54}$$

Typical values would be $n_1 \approx 1.5; \quad \Delta \approx 0.01$

which would give us $\Delta\tau \approx 50$ ns/km for a step-index fiber,

$$\approx \frac{1}{4}\text{ ns/km} \quad \text{for a parabolic index fiber.} \tag{4.55}$$

The small value of $\Delta\tau$ for a parabolic index fiber is due to the fact that the optical path lengths traversed by all the rays along the fiber are nearly the same, and hence they are characterized by the same transit time. On the other hand, in a step-index fiber, rays traveling at different angles are characterized by different transit times, leading to a higher value of $\Delta\tau$.

Thus, having a near parabolic variation of refractive index leads to considerable decrease in the intermodal dispersion. It is for this reason that near parabolic index optical fibers were used in the first- and second-generation optical communication systems during 1970s.

4.5.2 Waveguide dispersion

Even in a single-mode fiber, there is a temporal broadening of a pulse that is due to the explicit dependence of the normalized propagation constant b on V [see Eq. (4.31)]. This temporal broadening is given by (see e.g. Ref. 2):

$$\Delta\tau_w = -L\frac{n_2\Delta}{c}\left(\frac{\Delta\lambda}{\lambda}\right)V\frac{d^2}{dV^2}(bV) \tag{4.56}$$

where $\Delta\lambda$ represents the spectral width of the source; the subscript w refers to the fact that we are considering waveguide dispersion. For a step-index fiber, an accurate semi-empirical expression for $V(bV)''$ is given by [11]

$$V(bV)'' \approx 0.080 + 0.549\,(2.834 - V)^2; \quad 1.3 \le V \le 2.6 \tag{4.57}$$

Thus, for a step-index single-mode fiber with

$$n_1 = 1.47,\ \Delta = 0.0025, \text{ and } a = 4\ \mu m$$

we have

$$V \approx 2.0 \text{ for } \lambda = 1.3\ \mu m$$

and

$$V(bV)'' \approx 0.462$$

which gives for the waveguide dispersion coefficient

$$D_w \equiv \frac{\Delta\tau_w}{L\Delta\lambda} \approx -4.34 \text{ ps/km-nm}$$

The dispersion is measured in units of picoseconds per nanometer spectral width of the source per kilometer length of the fiber. Thus, for a laser diode having a spectral width $\Delta\lambda \approx 2$ nm, we will have

$$\Delta\tau_w = -8.68 \text{ ps/km}$$

We should mention here that the waveguide dispersion is due to the intrinsic characteristics of the waveguide and will occur even when the refractive indices of the core and cladding are strictly independent of the wavelength.

4.5.3 Material dispersion

In this section, we will briefly consider material dispersion, which is due to the fact that any light source has a certain spectral width $\Delta\lambda$ and different wavelength components travel with different group velocities. For a source characterized by spectral width $\Delta\lambda$, the pulse dispersion is given by (see e.g. Ref. 2 and 3)

$$\Delta\tau_m = -\frac{L}{c\lambda}\left(\lambda^2\frac{d^2n}{d\lambda^2}\right)\Delta\lambda$$

where λ represents the free space wavelength and the subscript m refers to the fact that we are considering material dispersion. Since dispersion is usually measured in picoseconds per kilometer length of the fiber per nanometer spectral width of the source, we take

$$L = 1 \text{ km} = 10^3 \text{ m}, \Delta\lambda = 1 \text{ nm} = 10^{-9} \text{ m}$$

and $c = 3 \times 10^8$ m/s to obtain the material dispersion coefficient

$$D_m \equiv \frac{\Delta\tau_m}{L\Delta\lambda} = -\frac{1}{300\lambda}\left(\lambda^2 \frac{d^2n}{d\lambda^2}\right) \text{ ps/km-nm}$$

where λ in the denominator is in meters, and the quantity inside the parenthesis is a dimensionless quantity. For example, for pure silica at $\lambda \approx 0.85\ \mu m = 0.85 \times 10^{-6}$ m

$$\left(\lambda^2 \frac{d^2n}{d\lambda^2}\right) \approx 0.0289 \Rightarrow D_m \approx 113 \text{ ps/km-nm}$$

For a laser diode of spectral width $\Delta\lambda \approx 2$ nm, the material dispersion would be

$$\Delta\tau_m \approx -226 \text{ ps/km}$$

On the other hand,

$$\left(\lambda^2 \frac{d^2n}{d\lambda^2}\right) \approx -0.010 \text{ at } \lambda \approx 1.55\ \mu m \quad \Rightarrow D_m \approx 22 \text{ ps/km-nm}$$

For a laser diode with $\Delta\lambda \approx 2$ nm, the material dispersion would be

$$\Delta\tau_m \approx 44 \text{ ps/km}$$

Notice that the waveguide and material dispersions are of opposite signs. Fiber designers can design the fiber refractive index profiles in a way so as to tailor the magnitude of its waveguide dispersion such that the waveguide and material dispersions cancel each other at a desired wavelength to yield negligible total dispersion (see e.g. Ref. 2).

References

1. A.W. Snyder and J.D. Love, *Optical Waveguide Theory*, Chapman & Hall, London (1983).
2. Ajoy Ghatak and K. Thyagarajan, *Introduction to Fiber Optics*, Cambridge University Press, Cambridge (1998). Reprinted by Foundation Books, New Delhi.
3. Ajoy Ghatak, *Optics*, 4th Edition, McGraw-Hill, New Delhi (2008).
4. D. Gloge, *Weakly Guiding Fibers*, Appl. Opt., vol. 10, 2252, (1971).
5. Ajoy Ghatak, A. Sharma and R. Tewari, *Fiber Optics on a PC*, Viva Books, New Delhi (1994).

6. H.D. Rudolph and E.G. Neumann, "Approximations for the eigen values of the fundamental modes of a step-index glass fiber waveguide", Z. Nachrichtentech., vol. 4, p. 328 (1976).
7. Ajoy Ghatak, I.C. Goyal and R. Varshney, *FIBER OPTICA: A software for characterizing fiber and integrated-optic waveguides*, Viva Books, New Delhi (1999).
8. D. Marcuse, "Loss analysis of single-mode fiber splices", Bell. Syst. Tech. J., vol. 56, p. 703 (1977).
9. A. Ankiewicz and C. Pask, "Geometric optics approach to light acceptance and propagation in graded index fibers", Opt. Quant. Electron., vol. 9, p. 87 (1977).
10. D. Gloge and E.A.J. Marcatili, "Multimode theory of graded-core fibers", Bell. Syst. Tech. J., vol. 52, p. 1566 (1973).
11. D. Marcuse, "Interdependence of waveguide and material dispersion", Applied Optics, vol.18, p. 2930 (1979).
12. Ajoy Ghatak and K. Thyagarajan, *Optical Fiber*, in Springer Handbook of Lasers and Optics (Ed: Frank Träger), Springer Science, New York (2007).

PART II

5

Fiber-end Preparation and Light Coupling

M.R. Shenoy and Sunil K. Khijwania

5.1 Introduction

In communication systems as well as sensing applications of optical fiber, the light guiding medium must be coupled with source and the detector. In general, there could be three types of couplings involved in designing such systems, namely, source-to-fiber, fiber-to-fiber, and fiber-to-detector. All such couplings are associated with power loss, which must be minimized in order to realize an efficient transmission system. Hence, the coupling efficiency is a critical design aspect for any such systems, and influences the over-all system performance. The basic requirements in order to ensure an efficient coupling are the proper (flat, smooth, normal to the propagation axis) fiber-end preparation and precise alignment, avoiding light deflection or scattering. This chapter describes fiber-end preparation and effective power launching techniques in detail.

5.2 Aim

Preparation of fiber-end face and launching light into an optical fiber.

5.3 Apparatus

Optical breadboard, laser source, laser source mount, microscope objective (20X), microscope objective holder, XYZ-translational stage, photodetector with multimeter, photodetector holder, two fiber chucks/V-grooves, two post-bases and 3 posts, 2 m length (approx.) of a single-mode as well as a multimode fiber, razor blade, fiber cutter, and index matching liquid.

5.4 Fiber-end preparation

The very first step that one must follow before making any measurement/application involving use of optical fibers is to obtain optically-flat fiber-end-faces perpendicular to the axis of the fiber, in order to ensure a good coupling efficiency between a light source and a fiber. This process is referred to as 'fiber-end preparation'. The acceptable optical quality of the end-finish and tolerance allowable with respect to end-face angle (i.e. the angle between the fiber-end-face and the plane perpendicular to the axis of the fiber) depend on particular measurement/ application. Conventional grinding and polishing techniques can indeed produce a very smooth end-face that is perpendicular to the fiber axis. In this process, the test fiber is bonded with an optical epoxy within a glass/fused silica capillary tube, and the end of the combined structure is ground and polished together. However, this method is quite tedious and time-consuming and is rarely used in practice except in very special cases, where one requires an extremely high quality fiber-end-face e.g. in fiber-based Fabry-Perot resonators, and high-quality end-faces of connectorized fiber pigtails and patch cords.

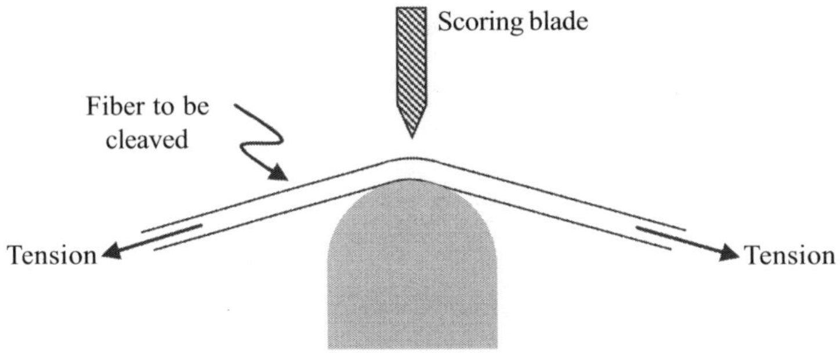

Fig. 5.1: Controlled-fracture of an optical fiber for its end-preparation through 'bend, score and break' technique.

Highly smooth and perpendicular end-face can be produced by means of a controlled-fracture technique, which is based on 'bend-score-and-break' method for cleaving fibers, and is sufficient for most experiments. In this method, after removal of the outer protective jacket, the fiber to be cleaved is first scored with a scriber to initiate a crack on its surface. The scriber, which is required to be harder than silica glass, is typically a diamond-edged or a tungsten carbide blade. The fiber is then bent over a curved surface under tension, as shown in Fig. 5.1. For a silica fiber of 125 μm overall diameter, the radius of curvature of the surface is about 57 mm. This action produces a stress distribution in the fiber cross-section. The maximum stress occurs at the scratch so that a crack starts to propagate through the fiber, which eventually leads to cleaving of the fiber when the fiber is gently pulled apart by hands [1].

A variety of fiber cutting tools based on this methodology have been developed and are being used extensively. Nevertheless, the curvature of the fiber and the amount of the tension required to induce cleavage should be precisely controlled in the controlled-fracture method. If the stress distribution across the fiber is not properly controlled, the fracture propagating across the fiber can diverge into several adjacent cracks. This forking produces surface defects such as a "lip" or "hackle" on the fiber-end, as shown in the Fig. 5.2.

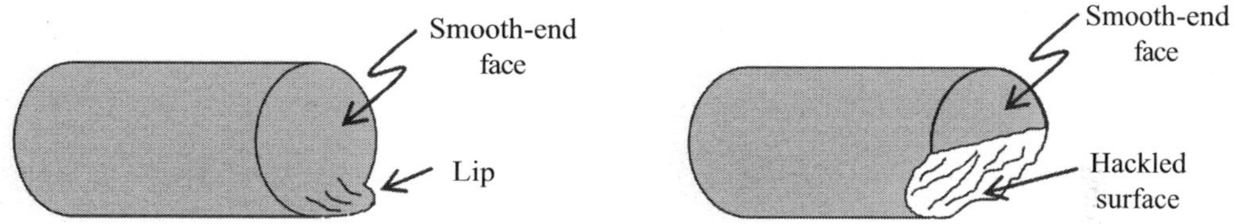

Fig. 5.2: Two examples of improperly cleaved fiber-ends.

In the absence of a commercial tool, one can obtain, with some experience, good quality fiber-end-faces by following the procedure described below:

1. Remove the outer protective jacket of the fiber over a few centimeters length near the fiber-end. Depending on the material used, the protective jacket may be removed using a suitable chemical solvent (e.g. dichloro-ethylene) or may be simply peeled off using a sharp razor blade.
2. The bare fiber is held by the thumb and the middle finger, and is positioned on the curved surface near the tip of the fore-finger, the palm facing upwards (see Fig. 5.3).
3. A small crack is then initiated by gently scoring on the surface of the fiber (which is positioned on the fore-finger) with a steady hand, in a direction perpendicular to the fiber length.
4. Now, without changing the grip on the fiber, the fiber length beyond the crack is gently pulled apart along the axis of the fiber.

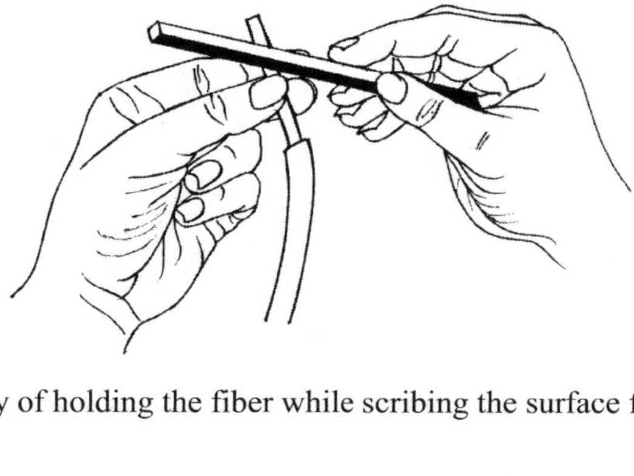

Fig. 5.3: The correct way of holding the fiber while scribing the surface for end preparation.

As a result, the fiber is likely to cleave with a good end-face. If this does not happen, then repeat the above procedure by initiating a fresh crack, at a slightly different position on the fiber.

A quick and effective assessment of the quality of the cleaved fiber-end-face can be made by launching light from the laser diode into the fiber at the farther end, and observing the output light on a screen placed a few centimeters away. Poor cleavage is revealed by the presence of scattered light in the form of a 'comet tail' stretching away from an otherwise circular shaped far-field spot on the screen. Further, if the fiber is poorly cleaved, the fiber-end can be seen glowing even when viewed from the sides, as a result of scattering [2]. In the case of single-mode fibers, one can also observe one or two bright and dark rings circumscribing the central bright spot on a screen placed near the fiber-end. The fiber-end will be good enough for most measurements if each of these rings are distinct, and appear to have uniform illumination all over the circumference.

5.5 Launching light into an optical fiber

The most common method of launching light into an optical fiber in a laboratory experiment involves use of a suitable microscope objective (MO) in between the source and the fiber. The choice of the MO depends on the type of source and the fiber under consideration. The two important parameters of a MO are the *magnification* and the *numerical aperture* (NA), usually inscribed on the metallic case of the MO. Normal MOs typically comprise of a combination of 3 lenses, while MOs with special imaging requirements may have a number of lenses forming a complex lens system. However, for our purpose, MO is nothing but a good quality lens to focus light (with minimum aberrations) from a collimated source into a fine spot at its focus, where the fiber-end is required to be approximately positioned in order to couple light into the fiber.

5.5.1 MO as a lens

Consider the normal incidence of a collimated beam of light on the input aperture of the MO, completely filling its aperture (see Fig. 5.4). If θ_a is the semi-angle of the focused cone of light

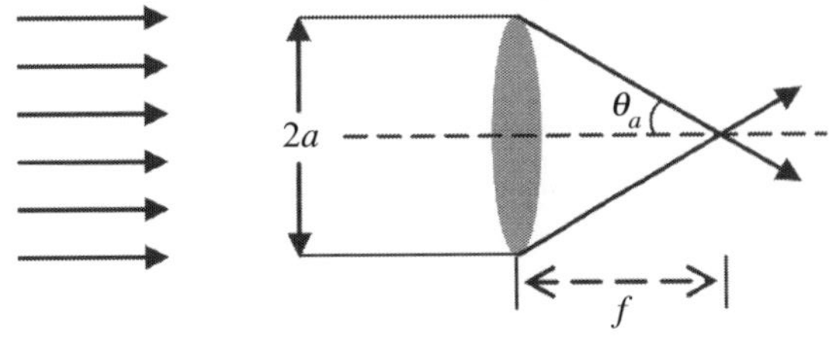

Fig. 5.4: A lens representing a MO of focal length f and numerical aperture sin θ_a.

(on the other side), then the numerical aperture (NA), by definition, is given by the value of a/f ($= \tan \theta_a \cong \sin \theta_a$). Thus, θ_a represents the maximum angle that the rays subtend with the axis of the MO at its focus. From Fig. 5.4, it is obvious that, for a given diameter of the front-end lens of the MO, larger NA implies smaller focal length and vice versa. This is an important observation for the user of a microscope objective for coupling light into an optical fiber, because the fiber-end should be placed in the close vicinity of the focus for achieving maximum light coupling into the fiber. The actual focal length of a MO corresponding to a given NA varies from one manufacturer to another and also upon the type of MO and the material of the lenses used. However, just to give the details for the MOs which could be used in the experiments described in this book, we have listed a few typical parameters:

Magnification	NA	focal length
10X	0.25	~10 mm
20X	0.40	~4 mm
40X	0.65	~2 mm

5.5.2 Aligning the source and the MO

In order to launch light into an optical fiber using an MO, the first step involves aligning the MO with the source; this is an important requirement for coupling maximum possible light into the fiber. Usually the source is either a laser or a tungsten halogen/Xenon-arc lamp. The source is mounted horizontally on a platform having provision for height and tilt adjustments. The MO is mounted on the microscope objective holder in front of the source at the same height as that of the source. The axis of source should be made exactly horizontal, and coinciding with the axis of MO. Provision of tilt-adjustment for MO holder will also be quite helpful.

In the case of a (collimated) laser source, e.g., He-Ne laser, one can see several spots in the back-reflected light due to reflection of the beam at different lens-surfaces in the MO. The relative positions of MO and the laser are adjusted so that almost all the back reflected spots coincide/overlap near the output aperture of the laser and the MO. In this position, the axis of the cone of light emanating from the MO becomes almost collinear with the laser beam. Indeed, almost perfect alignment of a He-Ne laser and an MO can be very conveniently demonstrated by mounting the MO holder and the laser on the movable riders of an optical bench.

A tungsten halogen lamp is the most commonly used 'white-light source' in measurements involving multimode fibers such as numerical aperture measurement and the near-field scanning technique to determine the refractive index profile. The lamp is usually housed in a ventilated (or air-cooled) metallic chamber fitted with a variable aperture at the output. For a given MO, the launch NA can be controlled using the variable aperture. The MO can be aligned with the source by observing the intensity distribution of the focused light on a screen placed at a suitable distance in front of the MO, i.e. the intensity distribution in a cross-section of the cone of light emanating from the MO. When this circular cross-section appears uniformly bright about the center, and is concentric with the circular 'image' of the output aperture of the MO (which also appears on the screen as a weakly illuminated circular disc), reasonably good alignment is achieved.

5.5.3 Coupling light into the fiber

After preparing good quality fiber-ends, the fiber is firmly clamped over the fiber chucks with magnets/suitable holders, such that only a small length (not more than 5 mm) of the free fiber-end projects out of the chuck. In many of the laboratory experiments, the fiber can even be held in a simple V-groove fiber holder by using some glue or modeling clay near the two ends of the groove. The fiber chuck/holder is then mounted on a XYZ-stack, i.e. a stack of micropositioners that can provide translational movement along the three perpendicular directions—X, Y, and Z. With the help of the translational movement screws, the fiber-end is precisely positioned near the focus of the MO to achieve maximum light coupling into the fiber. A schematic of such a coupling arrangement is shown in Fig. 5.5. In order to obtain a stable light-coupling arrangement, it is essential to avoid any possible relative motion (say, due to vibrations) between the MO and the fiber-end. The following step by step procedure would help in achieving maximum light coupling into the fiber:

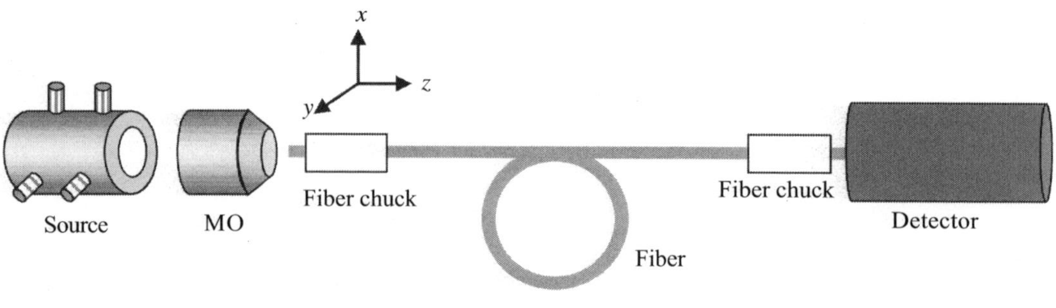

Fig. 5.5: Experimental set-up for launching light from a laser diode into an optical fiber using an MO.

1. Place the XYZ-stack such that the fiber chuck/V-groove is aligned with the MO. Adjust further to bring the fiber-end near to the focus of the MO.
2. Manipulate the X and Y (transverse) stages to get some light at the output end of the fiber. Maximize the output at this (longitudinal) position Z. Usually the coupling will be best, when the fiber-end is placed within the focus of the MO.
3. Now, move the fiber-end by a small distance in the +ve Z-direction (with respect to its position achieved in the step 2), and again maximize the output power using X and Y stages. If this maximum is lower than the previous maximum (at step no. 2), then move the fiber in the negative Z-direction and adjust for the maximum output power.

4. The process is repeated till the optimum fiber position corresponding to the best coupling efficiency is achieved.

5.6 Cladding mode stripping

Cladding modes may be defined as modes guided along the length of the fiber by the cladding-jacket or the cladding-air interface, in analogy to the definition of the guided modes. Here 'jacket' refers to the outer protective plastic/polymer layer, with which a bare fiber is coated immediately after it is drawn into a fiber from its preform. When light is launched into an optical fiber, the effective launch NA of the MO is usually chosen to be higher than the NA of the fiber. If the material of the jacket is of a lower refractive index than that of the cladding, then some of the rays transmitted through the core-cladding interface can satisfy the condition for total internal reflection at the cladding-jacket interface, giving rise to the excitation of cladding modes. Figure 5.6 illustrates this situation through the ray-optics. Note that the rays reflected by the cladding-air interface represent even higher NA than those which form the core modes and therefore, the bare fiber will also support cladding modes.

Since the jacket comprises of a coating of rapidly condensed fluid, the cladding-jacket interface is usually not uniform; further, the material of the jacket is relatively lossy. Thus, the behaviour of cladding-modes is quite unpredictable, and they attenuate rapidly as they propagate along the fiber. However, in many experiments on the characterization of an optical fiber, one uses a short length of the fiber, and therefore cladding modes may carry a significant amount of optical power to the detector, resulting in spurious signals. Thus, it becomes necessary, virtually in all experiments, to eliminate the cladding modes before they reach the detector.

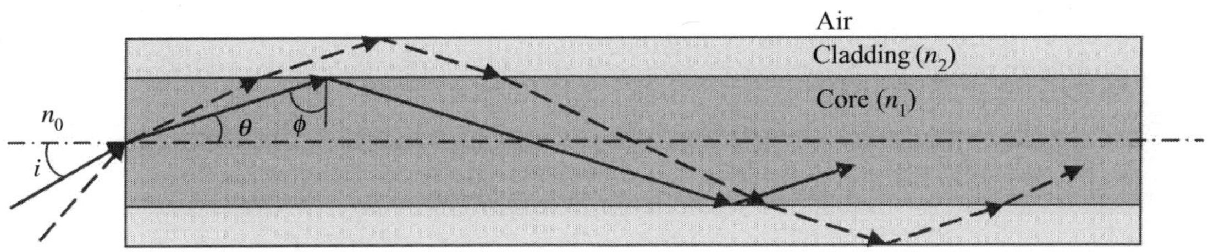

Fig. 5.6: Ray diagram illustrating the propagation of guided and cladding modes in an optical fiber.

A simple method of eliminating, or 'stripping off' the cladding modes involves applying an index-matching liquid such as glycerin or liquid paraffin over a few centimeters length of the bare fiber near both the input and the output ends. The liquid should, however, not come in contact with the fiber-end-faces. The refractive index of the liquid used should be slightly greater than or equal to that of the cladding, so that the rays transmitted through the core-cladding interface do not satisfy the condition for total internal reflection at the cladding-liquid interface. Thus, the rays entering the cladding will eventually leak out within a few centimeters length of the fiber.

5.7 Coupling of light from the fiber to a photodetector

Light from the output end of a fiber comes out in the form of a cone of semi-angle θ, which is determined by the numerical aperture of the fiber. In order to measure the output power in the visible and near infra-red region, a suitable detector such as Si photodiode is used. The fiber-to-detector light-coupling mechanism employed depends on the active area of the detector. In fiber characterization experiments, usually a large area detector (i.e. area ~1 cm^2) is used. In this case, the fiber output end is placed in front of the detector, such that the output from the fiber fills about 70% of the detector's active area, i.e. the output light-spot lies well within the active area of the detector. This ensures that the detector is not saturated locally and any variations in detector-surface response are averaged out. If the entire light spot lies within the active area, then the detector output is a measure of the total optical power incident on the detector. The detector may be connected to a power-meter, which shows photovoltage that is proportional to the photocurrent generated by the detector. Since the photocurrent is proportional to the optical power incident on the detector, the voltmeter reading is proportional to the optical power. Thus, one can measure the intensity distribution across an optical beam using this arrangement. The voltmeter readings typically vary from a few microvolts to a few millivolts, depending on the incident power.

The above situation corresponds to measurement of the total output power. In situations where the intensity distribution at the output is to be measured, a pinhole-masked large area detector is usually employed to scan the output light spot along a diameter. Now the detector output is proportional to the intensity of light at the position of the pinhole, and therefore the scan represents the intensity distribution.

5.8 Laser diode as an optical source

The principal light sources used for fiber optic communication applications are semiconductor laser diodes (also referred to as injection laser diodes) and light emitting diodes (LED). These devices are suitable for fiber transmission systems because they have adequate output power for a wide range of applications and their optical power output can be directly modulated by varying the input current to the device; they have a high efficiency and their dimensional characteristics are compatible with those of the optical fiber.

The major difference between LEDs and laser diodes is that the optical output from an LED is incoherent, whereas that from a laser diode is coherent. In a coherent source the optical energy is produced in an optical resonant cavity by the phenomenon of stimulated emission. The optical beam released from such a cavity usually has a very high degree of coherence, and is also very directional. The present day lasers are based on double heterostructures in which a thin active layer of a semiconductor with a lower bandgap is sandwiched between two larger bandgap semiconductors. The refractive index of the semiconductor decreases with increase in the bandgap. Thus, the refractive index of the central active layer is higher than the two surrounding regions. Such a refractive index profile can confine the emitted optical radiation to the active region by the mechanism of waveguidance, which occurs due to total internal reflection taking place at the boundaries [3]. Although the active region is very thin, this doesn't lead to a narrow slit-like output beam; there is substantial diffraction of the output at the interface with air. Thus, the output beam from a laser diode is not a fine spot or even a strip. It rapidly becomes a broad beam, and in order for light not to be wasted, a collecting lens should be placed immediately adjacent to the output. One can use collimating lenses to obtain a well-collimated output laser beam. The laser source usually used in the Educational Kits is a semiconductor laser fitted with an external collimating lens.

References

1. D. Gloge, P.W. Smith, D.L. Bisbee and E.L. Chinnock, "Optical Fiber-end Preparation for Low Loss Splices", *Bell Syst. Tech. J.,* vol. 52, pp. 1579–1588 (1973).
2. B.P. Pal, K. Thyagarajan and A. Kumar, "Characterization of Optical Fibers for Telecommunication and Sensors—Part I: Multimode Fibers" in *Fundamentals of Fiber Optics in Telecommunication and Sensor Systems,* B.P. Pal (Ed.), Wiley Eastern Ltd., New Delhi (1992).
3. A.K. Ghatak and K. Thyagarajan, *Introduction to Fiber Optics,* Cambridge University Press, Cambridge (1998).

6

Measurement of Numerical Aperture

M.R. Shenoy and Sunil K. Khijwania

6.1 Introduction

The numerical aperture (NA), as mentioned in Chapter 3, is a measure of the light collection efficiency of an optical fiber from an optical source, which is indeed a manifestation of the ability of confining this collected light inside the fiber through total internal reflections. In addition, it also helps in determination of the number of modes guided by a fiber. NA represents an important fiber characterization parameter, knowledge of which is required beforehand for different applications because specific application(s) require fiber(s) of different NA(s). It cannot be measured directly. In this chapter, we introduce the basic concept of NA and describe the corresponding measurement technique.

6.2 Aim

To determine the numerical aperture (NA) of a given multimode fiber from a measurement of its far-field.

6.3 Apparatus

Optical bread board, tungsten halogen lamp (THL), microscope objective (20X), microscope objective holder, XYZ-translational stage, pinhole photodetector with multimeter, photodetector holder, rotation stage, two fiber chucks, two post bases and 3 posts, approximately 2 m length of a multimode fiber, razor blade, fiber cutter, index matching liquid.

6.4 Theory

The extremely important parameter, which determines the maximum acceptance angle for the rays to be guided through an optical fiber when launched into the fiber from its input end is referred to as numerical aperture of the fiber and is given by (see Chapter 3)

$$NA = \sin i_m = \left(n_1^2 - n_2^2\right)^{\frac{1}{2}} \tag{6.1}$$

where n_1 and n_2 are the refractive indices of the core and the cladding, respectively, of the fiber. Figure 6.1 shows a particular meridional ray in a step-index multimode fiber. The light ray enters the fiber core from a medium of refractive index n_0 at an angle i with respect to the fiber axis and strikes the core cladding interface at a normal angle ϕ with respect to the normal to the interface. If ϕ is greater than the critical angle $\phi_c \left(= \sin^{-1}\left(\frac{n_2}{n_1}\right)\right)$, the ray is totally internally reflected and because of the symmetry of the structure, suffers repeated total internal reflections as shown in the Fig. 6.1. All such meridional rays follow zigzag paths along the fiber core, passing through the axis of the fiber after each reflection. The semi angle i_m of the acceptance cone for a step-index fiber is determined by this critical angle ϕ_c. Thus, if a ray is launched into the fiber at an angle $i < i_m$, it will be guided through the fiber, whereas a ray launched into the fiber at an angle $i > i_m$ undergoes only partial reflection at the core cladding interface and will eventually radiate completely into the cladding. Using a typical value of NA for a step-index multimode fiber to be 0.20, and assuming $n_1 = 1.45$, we get $i_m = 11.54°$.

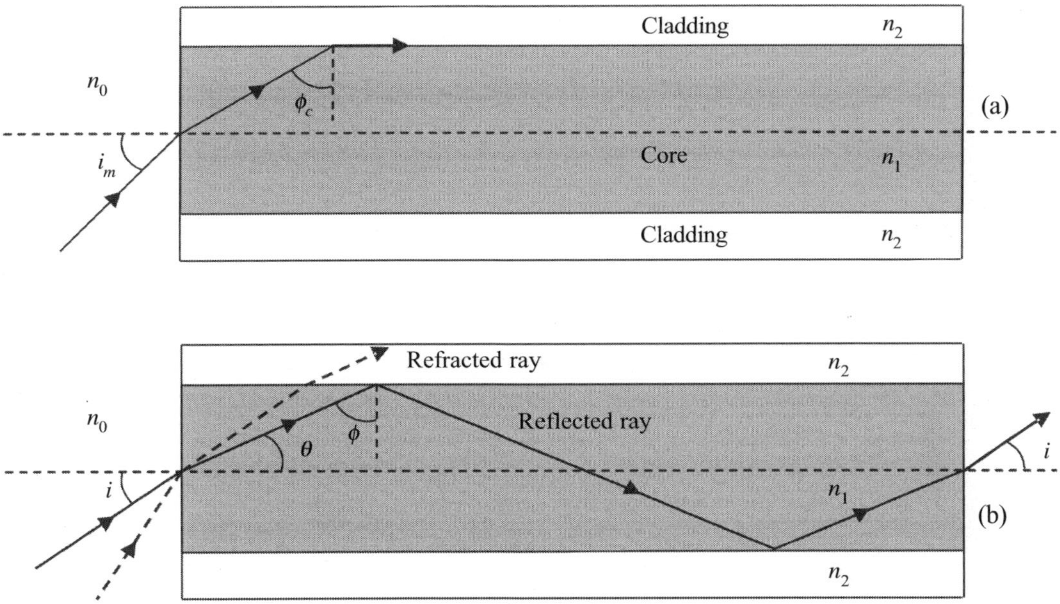

Fig. 6.1: A meridional ray entering a step-index optical fiber at an angle (a) i_m corresponding to critical angle incidence at the core-cladding interface; (b) $i < i_m$ gets guided through the fiber, and exits at the same angle i from the output end of the fiber. Rays incident at angles $>i_m$ will get refracted into the cladding, eventually loose all their energy into the cladding.

In a short length of straight fiber, ideally, a ray launched at an angle i at the input end should come out at the same angle i from the output end as shown in Fig. 6.1(b). Therefore far-field at the output end will also appear as a cone of semi-angle i_m emanating from the fiber end. Thus, the NA of the fiber is usually determined by measuring the angular dependance of the far-field of the fiber. The far-field of the fiber represents the field distribution at any distance $z >> (2a)^2/\lambda$ where a is the core radius of the fiber and λ is the operating wavelength.

The semi-angle corresponding to the 5% level of the far-field intensity maximum is denoted as the acceptance angle [1, 2]. This angle is usually small and hence, in the absence of a suitable arrangement for the measurement of the angular dependance of the far-field intensity, one may use direct visual estimation. In direct visual estimation, one carries out the measurements on the far-field projected on a plane perpendicular to the fiber axis by determining the diameter corresponding to the 5% intensity level (see Fig. 6.2) [1]. The numerical aperture is then estimated using the following relation:

$$NA = \sin i_m = \sin\left[\tan^{-1}\left(\frac{D}{2z}\right)\right] \tag{6.2}$$

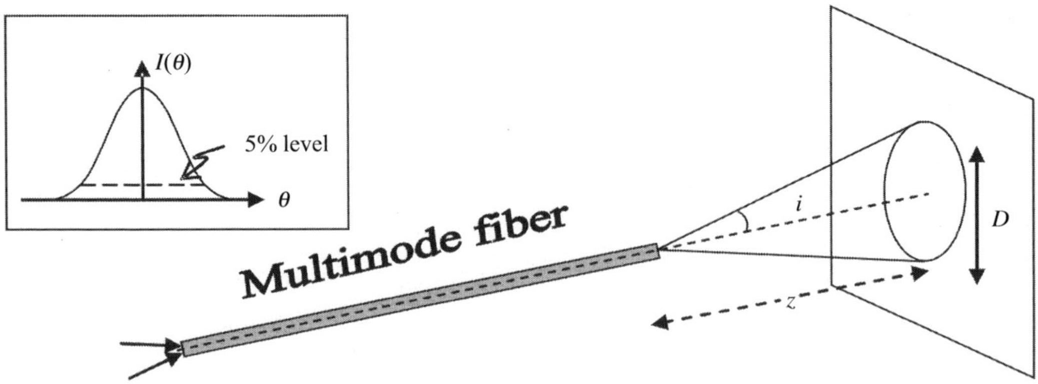

Fig. 6.2: Measurement of the diameter D of the spot, corresponding to the angle i at which the intensity drops down to 5% of the maximum value (at $i = 0$), on the screen placed at a distance z in the far-field from the output end of a multimode fiber can be used to estimate the NA of the fiber. Inset shows the far-field pattern and the 5% level.

6.5 Procedure

A. Scanning Method

Figure 6.3 shows the experimental set-up for the measurement of the numerical aperture of a multimode fiber. For this, one has to follow the following procedures step by step:

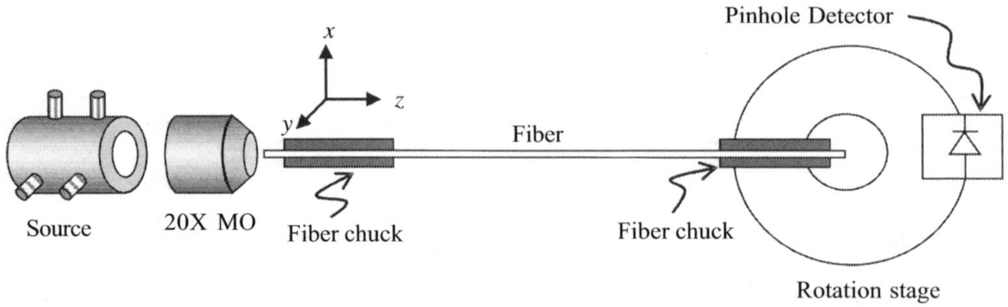

Fig. 6.3: Experimental set-up for scanning the far-field intensity distribution of a fiber.

1. Light from the tungsten halogen lamp (THL) is coupled into the given multimode fiber (about 2 m long) by using a 20X microscope objective (MO).
2. Fiber ends are prepared so that it has well-cleaved flat ends and then clamped over the fiber chucks with magnets. The cladding modes are removed by applying an index matching liquid (e.g. liquid paraffin) over a few centimeters of the bare fiber, near both the input and the output ends (after removal of the protective plastic jacket).

3. The output end of the fiber is clamped over a post base in such a way that the tip of the fiber gets positioned on the center of axis of rotation of the rotational stage (as shown in the figure). A set of cross wire placed over the mirror on the rotational stage helps in this placement of the fiber tip. The image of the fiber tip should coincide with image of center of the cross wire. For successful implementation of this step, one has to continuously look at the image of output end of the fiber on the mirror.
4. A pinhole-masked detector is mounted on the rotational stage and the height of the fiber tip is so adjusted that the pinhole is at the same horizontal level as that of the fiber end. The spatial resolution of detection is determined by the diameter of the pinhole, which may be about 0.2 mm.
5. Now, without disturbing the input coupling, far-field (circular) spot of the fiber is scanned in suitable steps (say, in steps of 0.2^{o}), and at each angular position, the detected power recorded by the detector is noted down.
6. Plot of the detector reading versus the angular position of the detector represents the far-field intensity distribution. From this plot the half angle i_m corresponding to 5% above the base line is determined.
7. Numerical aperture of the fiber is given by

$$NA = \sin i_m \tag{6.3}$$

The experiment may be repeated by slightly changing the input coupling.

B. Visual Method

A quick estimate of the NA is also possible by a simple visual method to determine the diameter of the far-field radiation spot as described in the following:

1. Several concentric circles of increasing radii – say, starting from 0.5 cm to 1.5 cm are drawn by a sharp pencil on a small paper screen placed normal to the fiber axis.
2. The screen is positioned in the far-field such that a line extending the axis of the fiber on the screen from its output end would have passed through the center of these circles, drawn on the screen.
3. The fiber end, which is mounted on a XYZ-stack, is moved slightly towards or away from the screen so that the emerging light spot exactly fills up one of the circles, lets say of diameter D (which is measured accurately).
4. The distance z between the fiber-end and the screen is measured accurately.
5. The NA is calculated using Eq. (6.2) for the direct visual estimation. Repeat the procedure for different values of z (and hence corresponding values for D), and find out the average value of NA. The value of NA thus obtained may be compared with the NA obtained by the scanning method.

6.6 Observations

S. No.	θ (degree)	Output Power	Normalized Output Power

6.7 Result

An experimentally measured angular distribution of the far-field output power for a given multimode fiber (NA = 0.26) is shown in Fig. 6.4, where the filled squares show experimentally observed values. The horizontal line drawn in the plot shows the level at which the intensity drops down to a value, which is 5% of the maximum output power. This line intersects with the plot at two angles, which are marked as θ_1 and θ_2 in Fig. 6.4.

The values of θ_1 and θ_2, as can be read from the above figure, are 182.1° and 209.3°, respectively. The angle i_m [= $(\theta_2 - \theta_1)/2$], can be calculated from these two values which comes out to be equal to 13.7°. The numerical aperture of the given multimode fiber is then calculated using Eq. (6.3), which is equal to 0.24. At the same time, the direct visual estimation method gives the

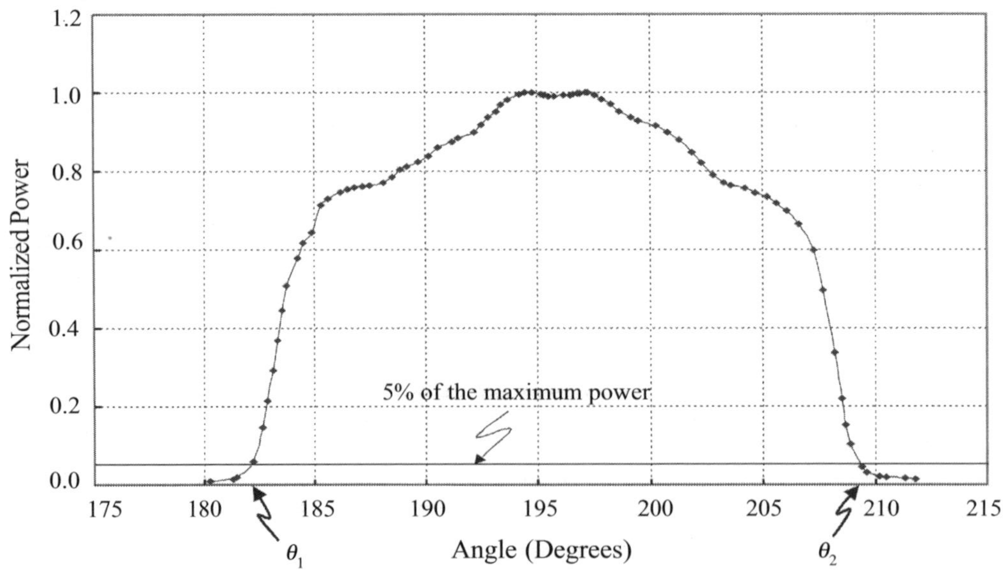

Fig. 6.4: Experimentally observed far-field pattern of a given step-index multimode fiber.

numerical aperture as equal to 0.20 for the same fiber, which is quite less than what we have measured using far-field technique. This shows that the far-field technique for the determination of numerical aperture is more accurate in comparison to the direct visual estimation method and the result obtained by this method is in reasonably good agreement with the actual value of the NA of the fiber.

6.8 Comments

In a multimode fiber, different mode groups suffer different attenuation rates, which are referred to as differential mode attenuation (DMA). NA measured by Eq. (6.3) is generally less than that estimated from Eq. (6.1) due to DMA.

In addition, NA is critically dependent on the excitation conditions. To ensure that all the guided modes are excited in a multimode fiber, an "overfill launch" is normally used at the input end i.e. one uses a microscope objective of NA higher than that of the fiber for launching light into the fiber. This results in the excitation of the cladding modes and the leaky modes, which quickly lose power as they propagate along the fiber. Further the higher order guided modes, which have a significant fraction of the total guided power in the cladding, attenuate faster as compared to the lower order modes. Therefore, in general, the modal power distribution within the fiber attains a steady state only after propagating some distance ($\approx$ a few tens of meters) along the length of the fiber. Consequently, the measured NA of a multimode fiber exhibits length dependence over relatively short length of the fiber.

The analyses discussed in the theory section are based on meridional rays. However, a greater power loss arises when skew rays are included in the analyses, since many of the skew rays that geometric optics predicts are trapped in the fiber are actually leaky rays. These leaky rays are only partially confined to the core of the circular optical fiber and attenuate as the light travels along the fiber. Thus, a detailed inclusion of skew rays will change the expression of the light acceptance ability, and hence NA of the fiber.

Although this method appears to be quite simple and straight forward, following points must be ensured at the time of the experiment:

1. The fiber end faces must be of good quality.
2. Cladding-mode strippers must be used.
3. The output end of the fiber must be positioned in such a way that the axis of rotation of the rotation stage passes through it.

6.9 Notes

1. In graded index fibers, the refractive index of the core is a function of the radial coordinate r, i.e., $n = n(r)$. Thus, the "local numerical aperture" at any point in the cross-section is defined as $\text{NA}(r) = \sqrt{n^2(r) - n_2^2}$. In general, the refractive index is maximum on the axis

of the fiber, and the value of NA in these fibers refers to the axial numerical aperture, namely $\sqrt{n^2(0) - n_2^2}$.

2. The method of determining the NA by a simple measurement on the far-field, described above, is not very suitable for single-mode fibers because of the significant diffraction effects from the small aperture, represented by the fiber-core at the end-face. However, the method can still be employed to obtain a rough estimate with some loss in accuracy.
3. The theoretical NA of a fiber can be estimated accurately from the fiber's refractive index profile, if available, using Eq. (6.1), and it is applicable to both single-mode and multimode fibers.

References

1. B.P. Pal, K. Thyagarajan and A. Kumar, "Characterization of Optical Fibers for Telecommunication and Sensors-Part I: Multimode Fibers" in *Fundamental of Fiber Optics in Telecommunication and Sensor Systems*, B.P. Pal (Ed), Wiley Eastern Ltd., New Delhi (1992).
2. A.K. Ghatak and K. Thyagarajan, *Introduction to Fiber Optics,* Cambridge University Press, Cambridge (1998).
3. A.W. Snyder and J.D. Love, *Optical Waveguide Theory*, Routledge Chapman and Hall, New York, (1984).

7

Mode Field Diameter of a Single-Mode Fiber

Arun Kumar and R.K. Varshney

7.1 Introduction

Typically about 20% of the guided light in a single-mode fiber is carried in its cladding in contrast to that in a multimode fiber, where almost all the guided light is confined within the core. This makes the core diameter just an auxiliary parameter for a single-mode fiber and what matters most is the radial extent of the guided LP_{01} mode in it. A parameter, which is used to quantify this radial extent, is known as the mode field diameter (MFD), which in practice is used to quantify physically the radial size of a single-mode fiber in place of its geometrical core size. MFD depends on the operating wavelength and fiber refractive index profile. For example, at shorter wavelengths, MFD is smaller than its magnitude at longer wavelengths because at longer wavelengths the V-number of the fiber would be smaller and hence the mode field would spread into the cladding to a greater extent. It is analogous to the core diameter in a multimode fiber indicating the net region of field confinement. It plays a key role in evaluating transmission characteristics of a single-mode optical fiber such as light coupling efficiency, splice/microbending loss and waveguide dispersion. Thus, MFD is a very important characteristic parameter of a single-mode fiber (SMF). This chapter introduces the basic concepts of mode field diameter and its measurement technique.

7.2 Aim

To determine the mode field diameter (MFD) of the fundamental mode in a given single-mode fiber (SMF) by a measurement of its far-field.

7.3 Apparatus

Bread board, laser diode/He-Ne laser, microscope objective (20X), microscope objective holder, XYZ-translational stage, pinhole masked photodetector connected to a multimeter, photodetector holder, rotation stage, two fiber chucks, two post bases and 3 posts, approximately 2 m length of a single-mode fiber, razor blade, fiber cutter, index matching liquid.

7.4 Theory

The guided modes supported by an optical fiber, at a given wavelength, have well defined intensity patterns which depend on various fiber parameters such as core-radius, core-cladding index difference and shape of the index profile. As a result, useful information about the fiber-parameters can be obtained by analyzing these intensity patterns, either in the *near-field* or in the *far-field* region. A *far-field* measurement determines the angular dependence of the output intensity, sufficiently far from the output end of the fiber; while a *near-field* measurement determines the spatial intensity distribution that exists at the plane of the output end of the fiber (see Fig. 7.1).

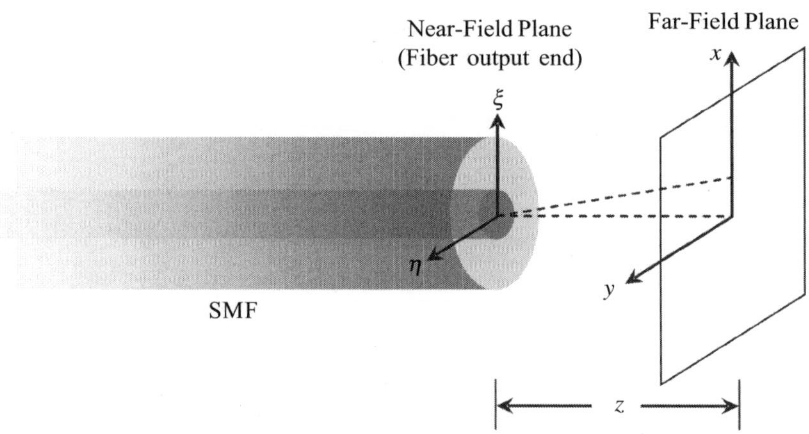

Fig. 7.1: Far-field geometry.

It may be mentioned here that the *far-field* pattern is nothing but the Fraunhoffer diffraction pattern of the *near-field*, and is usually said to start at a distance z_0 given by [1]

$$z_0 = \frac{(2a)^2}{\lambda} \tag{7.1}$$

where λ is the wavelength of the propagating light and a represents the core-radius of the fiber. In general, *far-field* measurements are made at a distance greater than 10 z_0 from the output end of the fiber. At $\lambda = 0.6328$ μm, z_0 is a few tens of micrometers for a typical single-mode fiber, and is few millimeters for a multimode fiber. Due to their small core dimensions and also since there are no optical components required at its output end, the measurement of *far-field* patterns is much more easier than *near-field* patterns of optical fibers.

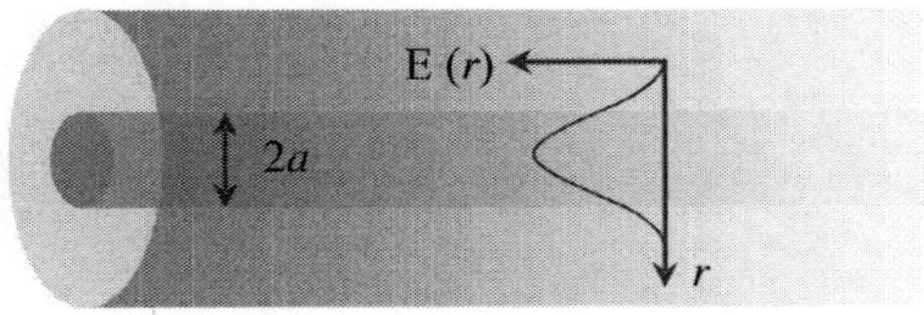

Fig. 7.2: Schematic diagram of the amplitude distribution of the propagating fundamental mode in a single-mode fiber.

Especially in single-mode fibers, it is the transverse field distribution of the propagating mode, rather than the core diameter and the numerical aperture, that is important in estimating several propagation and the performance characteristics of these fibers. Thus, *mode field diameter* (MFD), which is a measure of the transverse extent of the modal field distribution (i.e. of the LP_{01} mode), is an important parameter used to characterize a single-mode fiber. It is analogous to the core diameter in a multimode fiber indicating the region of field confinement, and it also takes into account the wavelength dependent field penetration into the fiber cladding. This parameter can be determined from the mode field distribution of the fundamental LP_{01} mode, which is shown in Fig. 7.2. It essentially specifies the transverse extent of this fundamental i.e. LP_{01} mode. Knowledge of MFD is very useful in estimating joint losses between two single-mode fibers, and also in calculating coupling efficiency, cutoff wavelength, backscattering characteristics, microbending losses, and even waveguide dispersion [2].

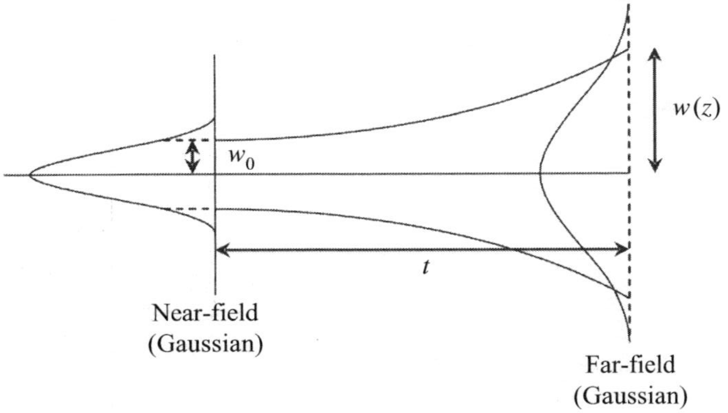

Fig. 7.3: Far-field pattern of a Gaussian beam. A Gaussian near-field has a far-field which is also Gaussian.

For most single-mode optical fibers of the type used in communication systems, the near-field of the fundamental mode can be well approximated by a Gaussian function of the form [3]:

$$\psi(r) = A \exp\left(-\frac{r^2}{w_0^2}\right) \tag{7.2}$$

where A is a constant. The quantity $2w_0$ gives the Gaussian MFD of the fiber. Theoretically, w_0 can be estimated by maximizing the launching efficiency between a free space Gaussian optical field and the single-mode fiber [4]. Marcuse has shown that for a step-index single-mode fiber, the following empirical expression may be used as a function of V for w_0 [4].

$$\frac{w_0}{a} \approx \left(0.65 + \frac{1.619}{V^{\frac{3}{2}}} + \frac{2.879}{V^6}\right); 0.8 < V < 2.5 \tag{7.3}$$

Since the far-field of a diffracting field, which is actually the Fraunhofer diffraction pattern, is Fourier transform of its near-field, it can be analytically shown that the far-field pattern of a Gaussian mode distribution [5] given by the above equation is again a Gaussian distribution (see Fig. 7.3) and the corresponding intensity pattern is given by

$$I(r) = B \exp\left[-\frac{2r^2}{w^2(z)}\right] \tag{7.4}$$

where B is a constant, and $w(z)$ is mode field radius (MFR) of the Gaussian modal field distribution in the far-field. Mode field radii w_0 and $w(z)$ are related through the following simple relation:

$$w^2(z) = w_0^2\left[1 + \frac{\lambda^2 z^2}{\pi^2 w_0^4}\right] \tag{7.5}$$

Thus, in the far-field the above equation can be approximated as

$$w(z) \approx \frac{\lambda z}{\pi w_0} \tag{7.6}$$

Accordingly the Gaussian MFD ($2w_0$) of a single-mode fiber can be easily obtained from a measurement of the angular distribution of its far-field as discussed below:

Using the above relation, the far-field intensity distribution of a Gaussian field is approximately given by [3]

$$I(r) = B \exp\left[-\frac{2\pi^2 r^2 w_0^2}{\lambda^2 z^2}\right] \tag{7.7}$$

$$= B \exp\left[-\frac{2\pi^2 w_0^{\,2}}{\lambda^2}\tan^2\theta\right]$$

where B is a constant, $\tan\theta = \dfrac{r}{z}$, θ being the far-field diffraction angle. The angle θ_e at which the far-field intensity drops down by a factor of e^2 from its maximum value ($\theta = 0$) (see Fig. 7.4) would then be given by [2, 3]

$$\tan\theta_e = \frac{\lambda}{\pi w_0}$$

which yields

$$w_0 = \frac{\lambda}{\pi \tan\theta_e} \tag{7.8}$$

Thus, by measuring θ_e one can easily calculate the Gaussian mode field diameter (MFD) $2w_0$.

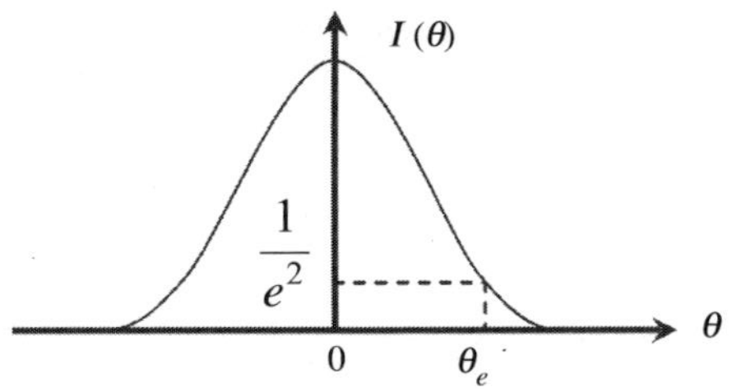

Fig. 7.4: Determination of θ_e from the measured angular distribution.

7.5 Procedure

Figure 7.5 shows the schematic of the set-up for the measurement of the mode field diameter of a single-mode fiber. For this, following procedures have to be followed step by step:

1. Mount the He-Ne laser in the aligner and adjust with the help of the aligning screws.
2. Fiber ends are prepared so that the fiber has well-cleaved ends. The cladding modes are removed by applying an index matching liquid (e.g. liquid paraffin) over a few centimeters near the input and output ends of the bare fiber. Fiber ends are then clamped over the fiber chucks with the clamping magnets.
3. Chuck carrying the output end of the fiber is clamped over a post base in such a way that the tip of the fiber is positioned on the axis of rotation of the rotational stage.

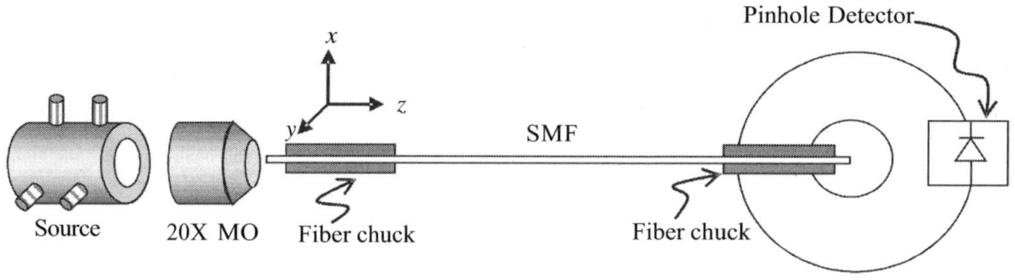

Fig. 7.5: Experimental set-up for scanning the far-field intensity distribution of a single-mode fiber.

4. Light from He-Ne laser is launched into one end of the given single-mode fiber (of about 2 meters length) using a 20X-microscope objective (as discussed in Chapter 5).
5. A pinhole-masked photodetector is mounted on the rotational stage. Height of the output fiber tip is so adjusted that the pinhole is positioned at the same horizontal level as that of the fiber end.
6. Now, without disturbing the input coupling, angular intensity distribution in the far-field (circular) spot of the fiber is scanned in suitable steps (say, in steps of 0.5°), and for each recorded angular position the multimeter reading is recorded. The measured data will correspond to the far-field angular intensity distribution.
7. Plot of the multimeter reading versus the angular position of the detector represents the recorded far-field intensity distribution. From this plot, calculate the value of θ_e at which the intensity is reduced to $\frac{1}{e^2}$ of the maximum intensity (at $\theta = 0$).
8. Using this angle, calculate the Gaussian mode field diameter ($2w_0$) using Eq. (7.8).

7.6 Observations

S. No.	θ-θ_{max} (degree)	Output power	Normalized Output power
1.			
2.			

7.7 Result

A typical variation of experimentally measured angular distribution of the far-field output power is shown in Fig. 7.6. This result corresponds to the given single-mode fiber (Fibercore, SM600).

In this figure, the filled squares show experimentally observed values, whereas the curve through the circles shows the Gaussian fit to these experimental data. The horizontal line drawn on the plot shows the level at which the intensity drops down to $\frac{1}{e^2}$ of the maximum output power. This line intersects the plot at two angles, which can be directly read from the plot. From this plot, θ_e (angle at which, the intensity drops down to $\frac{1}{e^2}$ of its maximum value) comes out to be $\approx 5.06^o$. Using this value of θ_e and the wavelength of the He-Ne laser (which is equal to 633 nm),

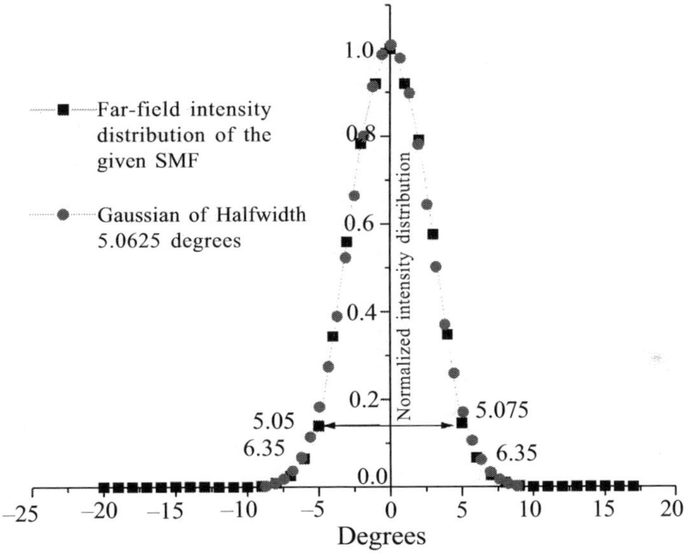

Fig. 7.6: A typical plot of far-field intensity distribution of a single-mode fiber.

Eq. (7.8) yields Gaussian mode field diameter (MFD) $\approx 4.5\ \mu m$. The best fitting Gaussian to the experimental data is also shown in the figure, which is in excellent agreement with the value provided by the manufacturer, i.e. 4.7 μm.

7.8 Comments

To achieve a good angular resolution, it is necessary that the far-field intensity pattern should be detected at a sufficiently large distance from the center of the fiber output end. Furthermore, the angular sector scanned in front of the fiber must be sufficiently wide (between $\pm 20^o$ and 25°) to completely include the main lobe of the radiation pattern.

Although this method appears to be quite simple and straightforward, following points must be ensured at the time of performing the experiment:

1. The fiber end faces must be of good quality.
2. Cladding-mode strippers must be used.
3. The output end of the fiber must be positioned in such a way that the axis of rotation of the rotation stage passes through the fiber tip.

The fiber used in this experiment was a Fibercore SM600, single-moded at 632.8 nm. If a single-mode fiber designed for operation at a given wavelength is operated at a longer wavelength, the modal field will spread more into the cladding of the fiber [3], thereby yield a relatively larger mode field diameter.

Finally, it should be noted that the method described here for the measurement of mode field diameter is not unique, and several other methods have also been proposed [3]. Further, one must also note that, in general, the mode field varies with variation in the refractive index profile; in fact it can deviate substantially from a Gaussian distribution in case of complex fiber refractive index profiles, for example, those encountered in dispersion shifted fibers like DSF and non-zero DSF (NZ-DSF) and dispersion compensating fibers (DCF). In such cases, most standard practice is to define the MFD of the fiber through the "Petermann MFD ($2w_p$)" [3, 6].

References

1. A.K. Ghatak and K. Thyagarajan, *Optical Electronics*, Cambridge University Press, Cambridge (1989).
2. E.G. Neumann, *Single-Mode Fibers: Fundamentals*, vol. 57, Springer-Verlag, Berlin (1988).
3. K. Thyagarajan, B.P. Pal and A. Kumar, "Characterization of Optical Fibers for Telecommunication and Sensors-Part II: Single-mode Fibers", *Int. J. of Optoelectronics*, vol. 3, p. 153 (1988).
4. D. Marcuse, "Loss analysis of single-mode fiber splices", *Bell Syst. Tech. J.*, vol. 56, p. 703 (1977).
5. B.P. Pal, *Fundamentals of Fiber Optics in Telecommunication and Sensor Systems*, New Age Publications, New Delhi (2002).

8 Refractive Index Profile of a Multimode Optical Fiber by the Near-Field Scanning Technique

Bishnu P. Pal

8.1 Introduction

The propagation characteristics of an optical fiber are almost completely determined by the refractive index profile (RIP) of the fiber, i.e., the refractive index distribution $n(r)$ across the fiber. For a given RIP, solution of the corresponding wave equation will yield modal propagation constants, mode fields, pulse dispersion characteristic etc. of the fiber. In multimode fibers, it has a significant influence on the bandwidth of the fiber; for example, in a multimode step-index fiber, typically, the length-bandwidth product is about 30 MHz-km, whereas in an optimally profiled graded-index fiber a figure as large as 18 GHz-km is feasible. Realizing this fact, manufacturers have made widespread efforts to fabricate high-bandwidth graded-core multimode fibers. This called for accurate, high-resolution index profile measurement techniques. In single-mode fibers, however, the index profile does not influence the bandwidth as much except when the fiber is operated near the *zero dispersion wavelengths* where the 'waveguide dispersion', whose value depends on the refractive index profile, could affect the propagation characteristics.

Out of the many different techniques to measure RIP of optical fibers reported in the literature, the transmitted near-field (TNF) scanning and the refracted near-field (RNF) scanning form the two most popular techniques; the former being relatively simple to implement. However, TNF technique is restricted to multimode fibers only. The RNF technique, on the other hand, is applicable to both single-mode and multimode fibers, and is presently the most widely used technique for precise index profile measurement. In the following, we describe the TNF technique to measure the refractive index profile of a graded-core multimode fiber.

8.2 Aim

To measure the near-field intensity profile of a multimode fiber and hence its refractive index profile.

8.3 Apparatus

Tungsten halogen lamp, assorted microscope objectives (20X, 40X), two XYZ-translation stages, two V-groove fiber mounts, a pinhole detector mounted on a translation stage with an output cable for connecting to a power meter, index matching liquid for use as cladding mode stripper.

8.4 Theory

The method essentially involves a scan of the near-field intensity distribution that exists at the output end of a short length (typically one to two meters) of the given multimode fiber, in which all possible guided modes are excited approximately equally by means of an incoherent Lambertian source, such as a tungsten halogen lamp (or an LED). A light source is called Lambertian if the power dP emitted from the plane surface of an incoherent source obeys the following:

$$dP = I_0 \cos\theta \, dA \, d\Omega \tag{8.1}$$

where $I_0 \cos\theta$ is the brightness of the Lambertian source[1], dA is an elemental area of the emitting surface, $d\Omega$ is the element of solid angle over which the radiation is emitted and θ defines the direction of emission (see Fig. 8.1).

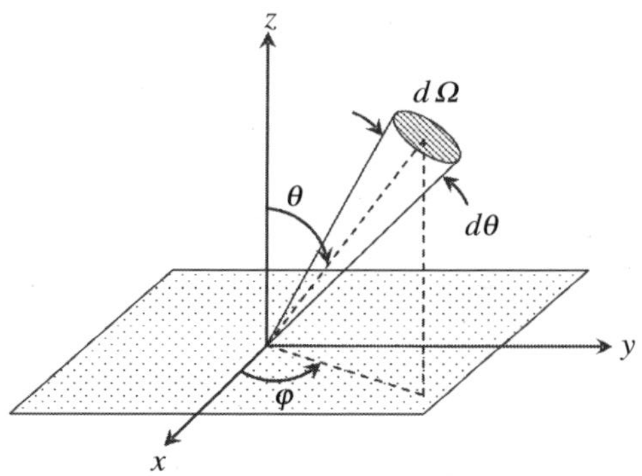

Fig. 8.1: Radiation emitted by a Lambertian source at an angle θ into an element of solid angle $d\Omega$.

[1] Output intensity of a Lambertian source is uniform in all directions. For example, polar diagram of output intensity as a function of emission angle for a perfectly Lambertian source will be a circle. In case of laser diodes or edge emitting LED's, output is more directional and intensity varies as $I_0(\cos\theta)^n$ with $n \geq 2$.

If the plane $z = 0$ is chosen to be the emitting surface of the source, then dP would be given by

$$dP = I_0 \cos\theta \sin\theta \, d\theta \, d\varphi \, dx \, dy \tag{8.2}$$

We assume that the light is launched into one end of the fiber from a Lambertian light source. On entering the fiber, the angle θ is transformed by refraction according to Snell's law

$$n_{\text{air}} \sin\theta = n(r)\sin\theta'(r) \tag{8.3}$$

where $n(r)$ is the local refractive index at a distance r from the axis and θ' is the corresponding local ray angle. In multimode fibers, where geometrical optics theory is valid, we know that the propagation constant β is related to the ray angle θ' inside the fiber core through

$$\beta = k_0 n(r) \cos\theta'(r) \tag{8.4}$$

Since the minimum value of β for a guided ray is $k_0 n_{cl}$, which is the plane wave propagation constant corresponding to the cladding, the maximum angle θ'_c (at position 'r') that a guided ray can make is given by

$$\cos\theta'_c(r) = \frac{n_{cl}}{n(r)} \tag{8.5}$$

The corresponding (maximum) angle at the fiber input face for this ray is given by (see Fig. 8.2)

$$n_{\text{air}} \sin\theta(r) = n(r) \sin\theta'_c(r) = \left[n^2(r) - n_{cl}^2\right]^{\frac{1}{2}} \tag{8.6}$$

which is the fiber's *local* numerical aperture (NA). Thus, at any point 'r' in the fiber core, only the radiation incident within the cone defined by its semi-angle $\theta_m(r)$ will be

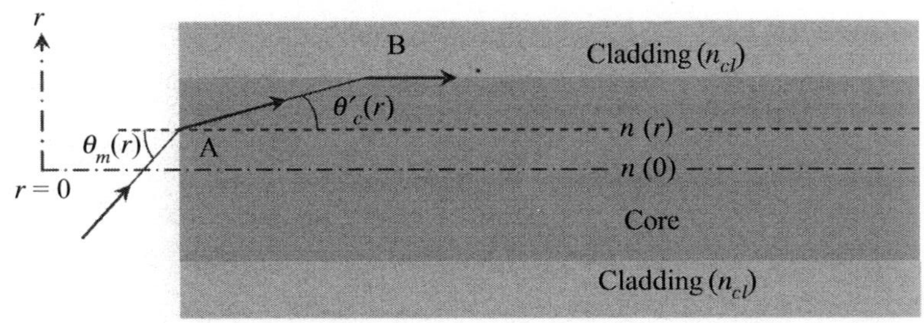

Fig. 8.2: Propagation of a ray having minimum value of β, incident at position r on a graded index fiber. [The ray path AB is shown to be straight line for clarity of the angle $\theta'_c(r)$].

accepted and guided by the fiber. If A_s denotes the area of the source, then the power launched into the cone defined by Eq. (8.5) is given by

$$P(r) = \int_0^{\theta_m} \int_0^{2\pi} I_0 \cos\theta \sin\theta \, d\theta \, d\phi \, A_s \tag{8.7}$$

where we have made use of Eq. (8.3). Since $n_{\text{air}} = 1.0$

$$\sin\theta \ \cos\theta \ d\theta = \frac{1}{2} d(\sin^2\theta) = \frac{1}{2} n^2(r) d(\sin^2\theta'(r)) \tag{8.8}$$

Thus,

$$P(r) = I_0 A_s \int_0^{2\pi} d\phi \int_0^{\sin^2\theta_m} \frac{1}{2} d(\sin^2\theta) = \pi I_0 A_s \left[n^2(r) - n_{cl}^2\right] \tag{8.9}$$

Equation (8.9) clearly shows that the power accepted at any point 'r' in the fiber core from a Lambertian source is directly proportional to square of the local numerical aperture. Evidently, the maximum power P_m that can be accepted by the fiber corresponds to the position r at which the fiber's NA is maximum. In an ideal fiber, this point would correspond to the fiber axis i.e. $r = 0$. However, many real fibers are characterized by an index dip at $r = 0$ so that the maximum index n_m does not necessarily occur on the axis. In terms of the maximum power $[P_m \equiv P(r = 0)]$ that can be coupled into a fiber, we obtain from Eq. (8.9)

$$\frac{P(r)}{P_m} = \frac{n^2(r) - n_{cl}^2}{n_m^2 - n_{cl}^2} \tag{8.10}$$

For small refractive index differences,

$$\frac{P(r)}{P_m} \approx \frac{n(r) - n_{cl}}{n_m - n_{cl}} \tag{8.11}$$

If all the guided modes propagate in the fiber without any differential attenuation or mode conversion, which essentially implies that we consider a short piece of fiber, which is relatively straight and devoid of any imperfections, then the same power distribution, which was launched at the input end, will exist at the fiber's exit face. Therefore, if the near-field of the given fiber is scanned by a small area photodetector, the detector output will yield the near-field power distribution of the fiber and hence RIP of the fiber as per Eq. (8.10). However, since the fiber core diameter is typically about 50 μm, it is difficult to directly scan near-field of the fiber with sufficient spatial resolution. In practice, therefore, one obtains a magnified image (of a convenient size) of the fiber output end, and this image is scanned by a small area photodetector [1].

Telecommunication-grade multimode fibers are designed to follow the nominal refractive index distribution given by

$$\begin{aligned} n(r) &= n_m\left[1 - 2\Delta\left(\frac{r}{a}\right)^q\right]^{\frac{1}{2}} && r < a \\ &= n_m[1 - 2\Delta]^{\frac{1}{2}} = n_{cl} && r \geq a \end{aligned} \tag{8.12}$$

where n_m is the axial refractive index at $r = 0$, a is the core radius and Δ is the relative core-cladding index difference

$$\Delta = \frac{n_m^2 - n_{cl}^2}{2n_m^2} \approx \frac{(n_m - n_{cl})}{n_m} \tag{8.13}$$

q is the index exponent which defines the shape of the core RIP. For example, $q = 1$ corresponds to a triangular core RIP, $q = 2$ corresponds to the parabolic core RIP and $q = \infty$ represents the step-index profile (see Fig. 8.3).

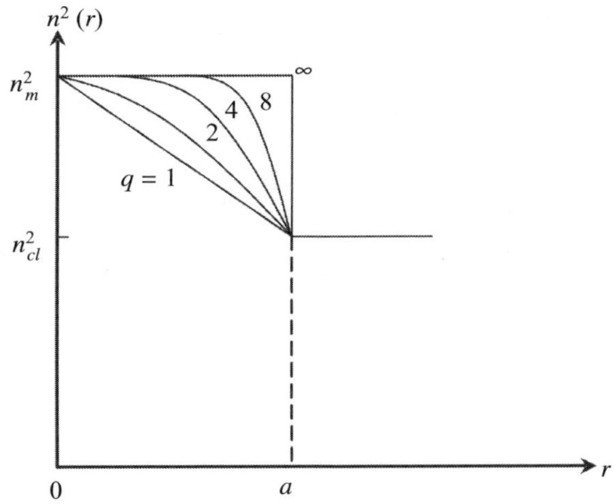

Fig. 8.3: Refractive index distributions of graded-core multimode fibers described by Eq. (8.12).

The so-called "optimum profile" corresponding to maximum transmission bandwidth occurs for $q \approx 2$. The exact value of q is determined by the material composition of the fiber [2].

Using Eq. (8.12) in Eq. (8.11) we get

$$\frac{P(r)}{P(0)} = 1 - \left(\frac{r}{a}\right)^q \tag{8.14}$$

or

$$\log\left[1 - \frac{P(r)}{P(0)}\right] = q \log\left(\frac{r}{a}\right) \tag{8.15}$$

Hence, a log-log plot of $\left[1 - \frac{P(r)}{P(0)}\right]$ against $\left(\frac{r}{a}\right)$ will yield a straight line of slope q and hence the profile shape.

It can be seen from Eq. (8.11) that a measurement of the near-field power distribution allows one to determine the refractive index difference $\left[n^2(r) - n_{cl}^2\right]$ except for the unknown constant $\left[n_m^2 - n_{cl}^2\right]$. However, $\left[n_m^2 - n_{cl}^2\right]$ is nothing but the square of the axial NA of the fiber, which can be determined independently. Further, if the value of the cladding index n_{cl} is known (which for high-silica fibers may be taken to be approximately 1.46 in the visible wavelength region, corresponding to that of fused silica[2]), absolute values of the indices $n(r)$ can be calculated from the measured near-field power distribution.

8.5 Procedure

Figure 8.4 shows the experimental set-up for the measurement of the refractive index profile of a multimode fiber. For this, one has to follow the following procedures step by step:

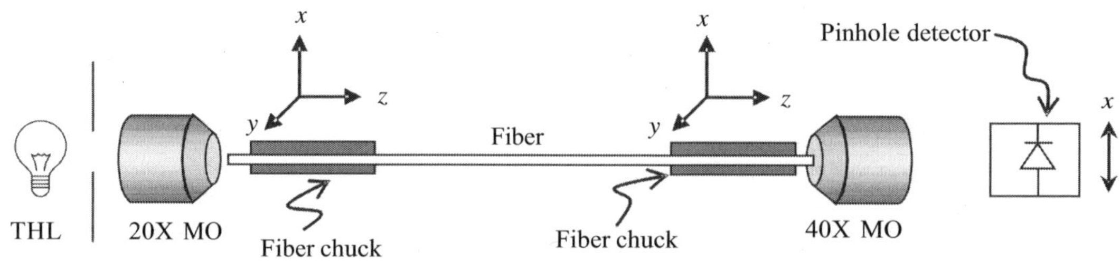

Fig. 8.4: Experimental set-up for near-field measurement of a multimode fiber.

1. Light from a tungsten halogen lamp (THL), which is an incoherent source of light, is coupled into the given multimode fiber (about 1 meter long) by means of a 20X microscope objective (MO). The NA of the objective should be equal to or greater than the fiber NA in order to ensure excitation of all the guided modes. Such a launch optics is known to effectively approximate a Lambertian source.
2. The fiber ends are mounted on two XYZ-translation stages. It is essential to have a good end-finish of the fiber ends by making good quality fiber cuts. Care must be taken to strip off the cladding modes by applying an index matching liquid, e.g. glycerin or liquid paraffin, over a few centimeters length of the bare fiber, near both its input and the output ends.
3. Using the 40X MO, a magnified image of the output end of the fiber is formed on the scanning plane of a pinhole detector. The pinhole diameter determines the spatial resolution of detection.

[2] At 1.55 μm, refractive index of fused silica is about 1.444.

4. The detector is mounted on a translation stage in order to scan the image along a diameter of the circular spot of the magnified near-field image.
5. To check that the near-field image is indeed formed on the detector plane, the following procedure may be followed: For a given position of the detector, the fiber exit face is moved to and fro (using micropositioners) along the axis of the optical system, in front of the imaging microscope objective. The position of the fiber for which the smallest and sharpest image could be seen on a screen coinciding with the plane of the detector will ensure the formation of the near-field image on the detector. This is easily understandable on the basis of the well-known lens equation: $\frac{1}{u}+\frac{1}{v}=\frac{1}{f}$ where u and v are the object (in this case, fiber output end) and image distances respectively, and f is the focal length of the imaging MO.
6. The detector output is connected to a power meter. The power meter readings can then be used to construct the near-field of the fiber by plotting power meter readings as a function of position of the detector.

Alternatively, if the detector is mounted on a motor driven translation stage or a scanner, the detector output could be directly fed to the Y-input of a XY-chart recorder. The near-field can then be obtained directly on the recorder as the 'X-arm' of the recorder is swept along the chart paper. The detector scanning speed and the sweep rate of the recorder should be chosen suitably to obtain a smooth record of the near-field profile.

8.6 Observations

S. No.	$\frac{P(r)}{P(0)} = X$	log $(1 - X)$	r (RHS of the axis)		r (LHS of the axis)	
			r/a	log (r/a)	r/a	log (r/a)
1. 2. . .						

8.7 Results

The near-field profile obtained as described above would be normally symmetric in shape about the central maxima (or minima, for fibers having an axial dip in its refractive index profile) corresponding to $\frac{r}{a}=0$. Since there is usually a rounding off at the core edge of the profile, it is difficult to precisely pinpoint the core-cladding interface on the measured near-field profile. This difficulty can be overcome by choosing the x-coordinate at which the near-field intensity drops down by 95% from its maximum value as the reference for $\frac{r}{a}=1$ (see Fig. 8.5, which shows the measured experimental result). A plot of $\log\left[1-\frac{P(r)}{P(0)}\right]$ against $\log\left(\frac{r}{a}\right)$ will yield a straight line of slope q, the index profile exponent.

In practice, it may be difficult to draw a single straight line connecting all the points due to the profile deviating from a perfect power-law distribution, and a mean value of q can only be obtained by estimating several localized q values corresponding to different small zones of the fiber core. Alternatively, a least square fit of Eq. (8.12) to the measured profile may be carried out to obtain the best-fitted value of q.

To determine the absolute values of refractive indices or the refractive index variation, we need to know the axial numerical aperture. The NA and hence Δ can be estimated from a measurement of the fiber's far-field. Removing the imaging microscope objective from the near-field measurement set-up and thereafter scanning the resultant output light from the fiber end would yield the far-field. Thus, from knowledge of n_{cl} and Δ, the measured near-field can be converted to a calibrated RIP with the help of Eq. (8.11).

As a final check of the estimated profile, the experimentally obtained values of q and Δ can be used in Eq. (8.12) and a comparison of the measured profile with that given by Eq. (8.12) can be made by plotting the latter also on the same graph.

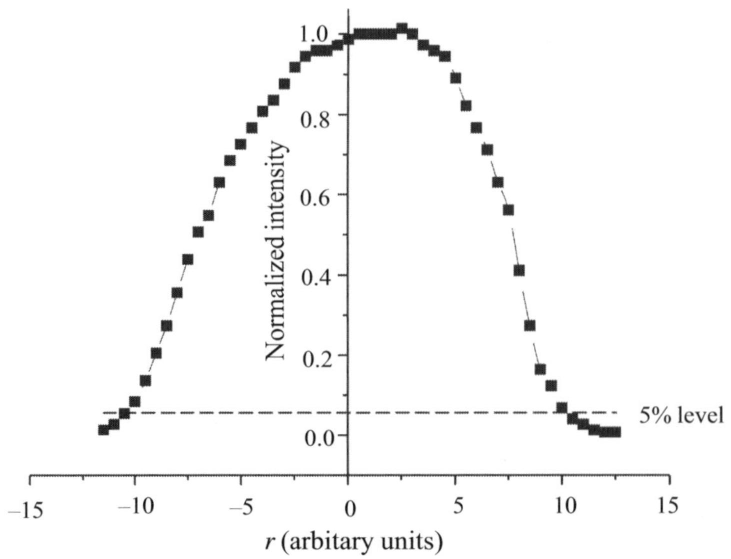

Fig. 8.5: Typical results of measurement of the near-field of a graded-core multimode fiber.

8.8 Comments

The most attractive feature of the method is that it provides a fast and relatively simple means to get a quick estimate of the core index profile and, for this reason alone, is employed by practically all fiber manufacturers. Nevertheless there is some controversy over the accuracy of the method. It is argued that since a relatively short fiber sample is excited in overfilled (both spatial and angular) launch conditions, the measured near-field may include a significant contribution from leaky rays. One may then be required to take account of this by incorporating a correction factor [3, 4] on the right-hand side of Eq. (8.11) in order to convert the measured near-field intensity profile to the fiber's refractive index profile. The presence of leaky rays leads to fiber length dependence of the measured near-field intensity distribution. This problem, however, is not as serious as it may seem because the leaky ray correction factors for a variety of profiles and fiber lengths were found to be almost identical up to a normalized radius $\frac{r}{a}$ of 0.8, and differed by roughly ±6% beyond $\frac{r}{a} = 0.8$ up to $\frac{r}{a} = 1\cdot$ Thus, in practice for a quick general estimate of the RIP, one ignores the presence of any leaky rays in the measured near-field profile of the fiber.

Since, the derivation of Eq. (8.11) is based on the assumption that all modes carry equal amounts of power, care should be taken to avoid (i) any unnecessary stress at the fiber mounts and (ii) sharp bends along the length of the fiber sample [5]. Care should also be taken to center the detector scan path on the fiber axis. Many laboratories use a vidicon with a monitor to scan the near-field. The near-field image of MCVD fibers is usually characterized by a dark central spot (corresponding to an axial index dip [5, 6]), which helps in defining the fiber center. This dark spot also helps the experimenter to check for optimal imaging of the fiber end-face on the detector plane by means of small adjustments in the (z-direction) position of the fiber end in front of the imaging objective so as to obtain maximum contrast in the imaged near-field pattern (NFP) at the plane of the scanning detector.

The inherent resolution afforded by the method is [5]

$$\delta r \sim \frac{\lambda}{\pi(\text{fiber } NA)} \tag{8.16}$$

For typical multimode fibers $NA \approx 0.2$ and with $\lambda \approx 0.65\ \mu m$, we get a maximum resolution of about 1 μm. On the other hand, detector spatial resolution is set by the size of the pinhole in front of it. If the fiber image size is l mm (diameter) at the detector plane and the pinhole diameter is d mm, then the number of independent measurement points on the image plane is roughly l/d; thus the resolution limit due to the pinhole is $Y = \left[\frac{2a}{l/d}\right]$ μm, a being the core radius in μm. For high measurement precision, Y should be less than or of the order of the theoretical resolution given by Eq. (8.16); this can be attained by having a relatively large image, produced by a combination of a microscope objective and an eyepiece [5]. However, in a graded-core fiber since the local fiber NA decreases from the core center towards the core-clad interface, there is a loss of resolution at the edges of the intensity profile [5, 6]; consequently, the core edges of the NFP are rounded (see Fig. 8.5). Another point which may be noted while implementing this method is that the emission spectrum of the optical source must be sufficiently broad, or the coherence length of the source should be very small, so that the finite number of guided modes do not interfere to produce a speckle pattern at the output end [7].

References

1. F.M.E. Sladen, D.N. Payne and M.J. Adams, "Determination of optical fiber refractive index profiles by a near-field scanning technique", Applied Physics Letters, vol. 28, p. 255 (1976).
2. B.P. Pal (Ed.), *Fundamentals of Fiber Optics in Telecommunication and Sensor Systems,* Wiley Eastern, New Delhi (1992).
3. M.J. Adams, D.N. Payne and F.M.E. Sladen, "Correction factor for the determination of optical fiber refractive index profiles by NFS", Electronics Letters, vol. 12, p. 281 (1976).
4. D. Marcuse, *Principles of Optical Fiber Measurements*, Academic Press, New York (1981).
5. B.P. Pal, "Refractive Index Profile Measurements in Optical Fibers", Electronics Research Laboratory Report STF 44 A 77208, Trondheim, Norway (1977).
6. B.P. Pal, K. Thyagarajan and A. Kumar, "Characterization of Optical Fibers for Telecommunication and Sensors-Part I: Multimode Fibers", *Int. J. of Optoelectronics*, vol. 3, p. 45 (1988).
7. M.J. Adams, D.N. Payne, F.M.E. Sladen and A. Hartog, Proc. Third ECOC, Munich (Berlin: VDE Verlag), p. 25 (1977).

9

Bend-Induced Loss in a Single-Mode Fiber

M.R. Shenoy

9.1 Introduction

Attenuation, which is a measure of the loss of optical power as light travels through the fiber, is one of the most important characteristics of an optical fiber. It is one of those fiber parameters, which restricts the transmission reach of signals in an optical fiber network. The main sources of signal power loss in a fiber can be broadly classified as (a) intrinsic, such as absorption, scattering etc., and (b) extrinsic, caused by two external mechanisms – macrobending and microbending. On the other hand, the bending loss could be gainfully exploited to realize specific attenuators and higher order mode filters. Thus, a pre-assessment of the loss characteristics of a fiber is important from both points of views mentioned above. In this chapter, we shall focus on the first of the two sources of extrinsic loss, namely macrobending and present an experimental technique for its measurement. Second source of extrinsic transmission loss (i.e. microbending) would be covered in Chapter 11.

9.2 Aim

To study bend-induced loss in a single-mode fiber.

9.3 Apparatus

Optical bread board, laser diode, laser diode aligner, microscope objective (20X), microscope objective holder, XYZ-translational stage, photodetector with multimeter, photodetector holder, two fiber chucks, two post bases and 3 posts, a single-mode fiber (approx. 3 m length), brass rods with circular cross-section of different diameters (say, from 0.8 cm to 4 cm), razor blade, scotch tape, fiber cutter and index matching liquid.

9.4 Theory

Radiative losses occur whenever an optical fiber is subjected to extrinsic perturbations like bend of a finite radius of curvature [1-6], and such losses are called *bend loss*. An optical fiber can be subjected to two types of bends: (a) random microscopic bends of the fiber axis, for example, those that may occur when a fiber is sandwiched between two sand-papers, and (b) macroscopic bends having radii that are large compared to the fiber diameter, for example, those that may occur during cabling and laying of the fiber cable, respectively. Here, we are mainly concerned with the power radiated from a single-mode fiber due to the latter type of bending, which is sometimes referred to as *macrobending*. Any simple experiment that involves launching a laser light (e.g. from a laser diode) into a fiber that is first laid straight and then bent into an arc of a circle, would reveal that the fiber suffers radiation loss from the side at bends/curves along its path. This can be measured by measuring the drop in transmitted power with an optical power meter placed at the output end of the fiber – once laid straight, and then after introducing a bend in its lay.

For relatively small bends, this loss is extremely small and is essentially unobservable. As the radius of curvature decreases, *transmitted power* decreases substantially until at a certain critical radius of curvature, the loss i.e. drop in transmitted power becomes distinctly clear. If the bend radius is reduced below this threshold point, the bend-induced loss suddenly becomes extremely large.

We first look at the physical process involved in the bending loss. For simulating bend loss, one can consider a straight section of the fiber having been jointed to a curved section of the fiber, end of which is once again joined with another straight section of the fiber, as shown in Fig. 9.1. According to ray analysis, there are no truly bound rays in a bent optical fiber. Thus, every ray is leaky and radiates power through the mechanism of tunneling or refraction. For example, the *meridional rays* in a bent fiber are either tunneling rays or refracting rays, and *skew rays* lose power at successive reflections either through tunneling or refraction.

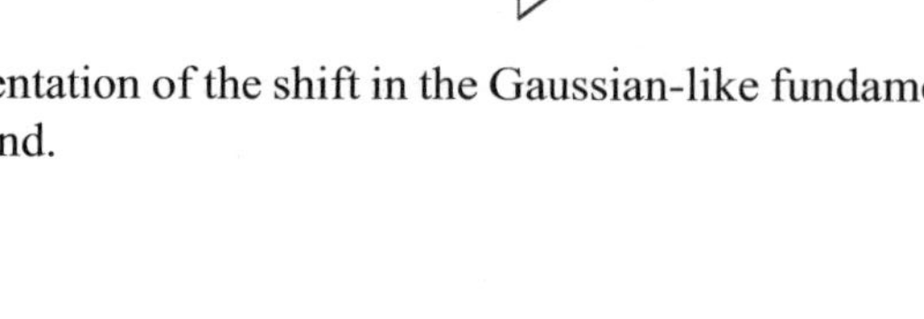

Fig. 9.1: A qualitative representation of the shift in the Gaussian-like fundamental mode away from the fiber axis at a bend.

Qualitatively, the bending loss can be explained by examining the modal electric field distribution as shown in Figs. 9.1 and 9.2. The bending loss can be accounted for by the *transition losses* and the *macrobend losses*. Transition loss is due to the abrupt change in curvature occurring, e.g., at the cross-sectional plane AA′ of the fiber shown in Fig. 9.1. The predominant effect of curvature on the fundamental mode is to shift the peak of the field distribution radially outwards (in the plane of the bend) by a distance r_d from the fiber axis as shown in Fig. 9.2. This shift, which is otherwise difficult to obtain exactly is given within the Gaussian approximation of the field as [2]

$$r_d = \frac{\beta^2 w_0^4}{2R_c} \tag{9.1}$$

where β is the propagation constant of the mode in the straight fiber, R_c is radius of curvature of the bend and w_0 is the Gaussian spot-size [6]. Thus the transition loss is due to the mismatch between the fields of the straight section and the offset field of the curved section over the transition plane AA′.

To get a feel for the numbers involved, let us consider a typical SMF with $n_1 = 1.46$, $n_2 = 1.457$ at an operating wavelength (λ_0) of 649 nm, and core radius $a = 2.1\ \mu m$, which yields $V \cong 1.90$. The corresponding Gaussian spot-size is calculated using the empirical relation [6]

$$w_0 = a\left(0.65 + \frac{1.619}{V^{\frac{3}{2}}} + \frac{2.879}{V^6}\right)$$

For the above SMF, $w_0 = 2.791\ \mu m$. If we further assume that $\beta \approx \frac{2\pi}{\lambda_0}\left(\frac{n_1 + n_2}{2}\right)$, then for $R_c = 0.5$ cm, one gets $r_d = 1.206\ \mu m$. The transition loss is given by [6]

$$\alpha_t(\text{dB}) = 4.34\left(\frac{r_d}{w_0}\right)^2$$

and the corresponding transition loss turns out to be 0.81 dB.

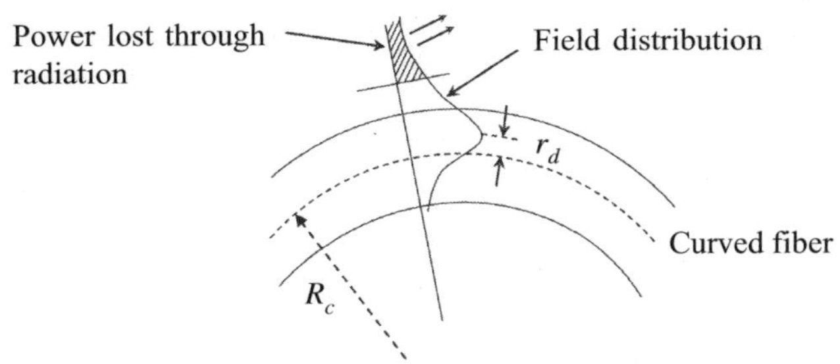

Fig. 9.2: Schematic representation of the bending loss in a section of fiber bent into an arc of radius R_c.

The second mechanism of loss is the actual transmission loss suffered due to radiation from side of the bent fiber. We know that any bound (core) mode has an evanescent *tail* in the cladding, which decays almost exponentially with distance from the core-cladding interface [6]. Since the evanescent tail moves along with the field in the core, part of the energy of a propagating mode travels in the fiber cladding. In a straight fiber of arbitrary profile, the modal field at every point in the cross-section propagates parallel to the fiber axis with the same phase velocity, so that the planes of constant phase are orthogonal to the axis. However, if the fiber is bent into a planar arc of constant radius, as shown in Fig. 9.2, it can be intuitively seen that the fields and phase fronts rotate about the center of curvature of the bend with constant angular velocity. Consequently, the phase velocity parallel to the fiber axis must increase linearly with the distance from the center of the curvature of the bend. Thus, the evanescent tail on the far side of the center of curvature must move faster to keep up in phase with the field in the core. At a certain critical distance d_c from the center of the fiber, the portion of the evanescent tail (in the cladding) would have to move faster than the speed of light in the fiber cladding to keep up with the core field. Since this is not possible as per fundamental laws of nature namely constancy of speed of light in a given medium, the optical energy in the evanescent tail beyond d_c radiates away [3]. Close to and within the core, the fields are accurately described by a local mode, provided the radius of curvature is sufficiently large.

The pure bend loss coefficient (in dB per unit length of the bent fiber) in a single-mode step-index fiber is given by [1]

$$\alpha_b = 4.34\left(\frac{\pi}{4aR_c}\right)^{\frac{1}{2}}\left(\frac{U}{VK_1(W)}\right)^2 \frac{1}{W^{\frac{3}{2}}}\exp\left[-\frac{4}{3}\frac{R_c}{a}\frac{W^3\Delta}{V^2}\right] \tag{9.2}$$

where, a is the core radius of the fiber, R_c is the radius of curvature of the bent fiber, $K_1(x)$ is the modified Bessel function of second kind. The parameters U, W, V and Δ are defined through

$$U = a\sqrt{k_0^2 n_1^2 - \beta^2}$$

$$W = a\sqrt{\beta^2 - k_0^2 n_2^2}$$

$$V^2 = U^2 + W^2$$

$$\Delta \cong \frac{(n_1 - n_2)}{n_2}$$

and

$$k_0 = \frac{2\pi}{\lambda_0}$$

It should be noted that in evaluating the above expression, the quantities λ_0, a, and R_c must have the same units, e.g., centimeters (cm). For the fiber parameters considered above, one gets

$V \cong 1.90$, $U \cong 1.49$, $W \cong 1.18$, $K_1(W) = 0.45$, which yields $\alpha = 51.52$ dB/cm when $R_c = 0.5$ cm. One can also calculate the transition loss, which is usually much smaller than the pure bend loss.

In the experiment, a few turns are given to the fiber so that the loss is measurable and then α_b is calculated as follows:

$$\alpha_b = -\frac{10}{L}\log\frac{P_2}{P_1} \tag{9.3}$$

where L is the length of the fiber (within the bend) i.e. $2\pi R \times$ (no. of turns), P_1 is the output power without the bend in the fiber and P_2 is the output power with the bend in the fiber.

9.5 Procedure

Figure 9.3 shows the schematic of the set-up for the measurement of the macrobending loss in a single-mode fiber. For this, following procedure may be followed step by step:

1. Mount the laser diode in the aligner and adjust with the help of the aligning screws.
2. Ends of a single-mode fiber of approximately 3 m in length are prepared so that it has well-cleaved ends.
3. The cladding modes are removed by applying an index matching liquid (e.g. liquid paraffin) over a few centimeters of the bare fiber, near the input as well as the output ends. Fiber ends are then clamped over the fiber chucks/holders. One end of the fiber is held in the XYZ-translational stage and the other in a V-groove that is mounted on a post base.

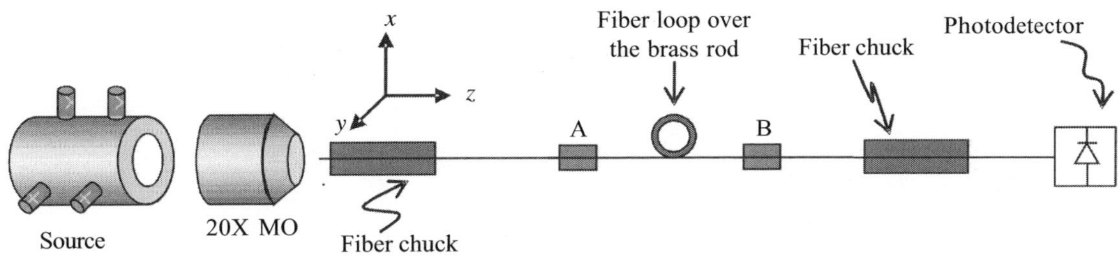

Fig. 9.3: Experimental set-up to study the macrobending loss in a single-mode fiber.

4. Light is launched from the laser diode using a 20X-microscope objective into the fiber.
5. The output end of the fiber is coupled to the photodetector, connected to the multimeter to measure the amount of light transmitted through the fiber. The multimeter reading is then noted down. This is taken as P_1 i.e. the power without the bending around the brass/aluminum rod. To measure P_2, i.e. the output power from the fiber after it is wrapped around the brass rod, following steps have to be carefully followed:

6. Fiber is clamped with cellotape at a distance 1 m from the input end (as shown in Fig. 9.3 at A) as well as just before the output end (as shown in Fig. 9.3 at B). In this way we would be having approximately 2 m length of the fiber in between the clamped points A and B.
7. Between the clamped points, a smooth brass/aluminium rod of uniform cross-section is placed as shown in the set-up and fiber is wound over it. For the brass rods of diameters between 2 to 4 cm, 10 to 20 turns (or more, to achieve significant difference between P_1 and P_2) of the fiber are wound. For the brass rods having diameters from 1 to 2 cm, you may choose suitable number of turns between 1 and 10. Between A and B, the fiber segments on either side of the rod should be kept stretched straight, bearing the same tension.
8. Vary the bending loss by changing the rods of different diameters and winding the fiber with suitable number of turns over it. Each time, note down the multimeter reading corresponding to the output of the detector.
9. Calculate the loss in dB per unit length using Eq. 9.3 and plot the measured loss as a function of the bending radius.

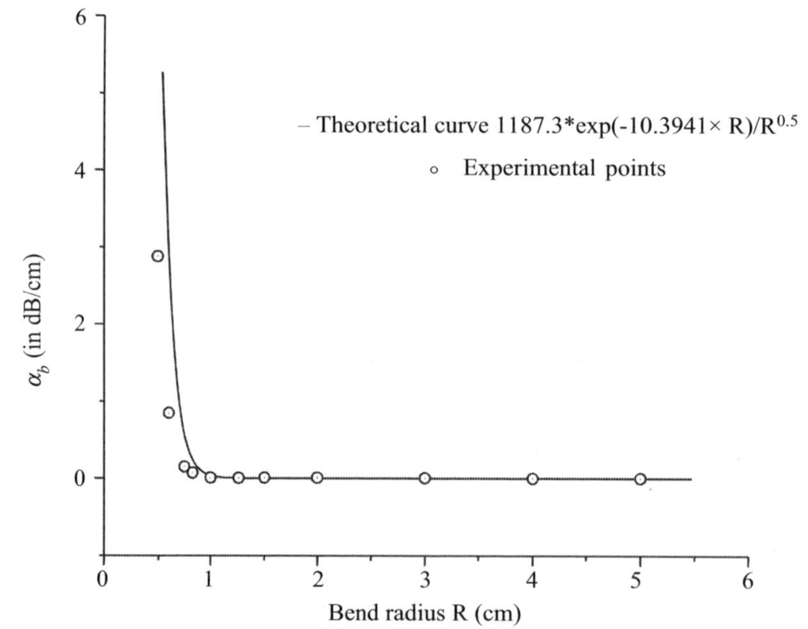

Fig. 9.4: Bend loss of a single-mode step-index fiber as a function of bend radius. The fiber core diameter ≈ 4.2 *μm*, $\Delta \approx 0.002$ and $V \approx 2.31$ at 0.633 *μm*; points represent the measured data and the continuous line represents the theoretical curve, obtained using Eqs. (9.2) and (9.3).

9.6 Observations

Input power without any bend =

S. No.	Radius of the rod, R	No. of turns	Total length of the bent fiber, L	Output power	Power loss (dB)	Bend-loss coefficient (α_b)
1.						
2.						

9.7 Result

A typical variation of experimentally observed bend-loss per unit length for a single-mode fiber (Fibercore, SM600) is shown in Fig. 9.4. The single-mode step-index fiber used for the experiment was having a nominal core diameter $\approx 4.2\ \mu m$, $\Delta \approx 0.002$, $V \approx 2.31$ at the He-Ne laser wavelength ($0.633\ \mu m$). In this figure, the circles show experimentally observed values, whereas the continuous curve shows the best fit to these experimental data. As can be observed, attenuation remains steady over a wide range of applied macrobend. This is the region where the loss due to scattering, absorption and other loss mechanism, excluding the macrobending dominates. Further, the macrobend loss increases as the macrobend radius of curvature decreases. Figure 9.4 also shows the theoretically simulated plot (obtained using Eq. 9.3), which is in excellent agreement with the experimentally observed data.

9.8 Comments

The experimental data usually starts deviating from the theoretical plot, which may be attributed to the non-inclusion of transition loss. Further, the procedure followed above for the bending loss measurement would not be suitable for multimode fibers since the bending loss per unit length in the fiber will be length dependent. In addition, one must be careful about the operating wavelength while evaluating the bending loss for SMF. Macrobending loss in a single-mode fiber depends on the mode field diameter (MFD). It increases with MFD. As MFD is a function of operating wavelength, macrobending loss also becomes a function of operating wavelength and increases with it. This is because of the fact that the evanescent tail of the fiber mode spreads more and more into the cladding with increase in wavelength and hence it becomes more susceptible to the macrobending.

References

1. A.W. Snyder and J.D. Love, *Optical Waveguide Theory*, Chapman and Hall, London, (1983).
2. W.A. Gambling and H. Matsumura, *Trans. IECE Japan* E 61, p. 196 (1978).
3. D. Marcuse, "Curvature loss formula for optical fibers", *J. Opt. Soc. Amer.*, vol. 23, p. 216 (1976).
4. D. Gloge, "Bending loss in multimode fibers with graded and ungraded core index", *Appl. Opt.*, vol. 11, p. 2506 (1972).
5. S.J. Garth, "Fields in bent single-mode fiber", *Electron. Lett.*, vol. 23, p. 373 (1987).
6. See, e.g., Ajoy Ghatak and K. Thyagarajan, *Introduction to Fiber Optics,* Cambridge University Press, Cambridge (1998).

10

Fiber-to-Fiber (Multimode) Splice Loss

Bishnu P. Pal and Varghese Paulose

10.1 Introduction

In any optical fiber telecommunication link, one or more splices/joints in the fiber cable are inevitable. The predominant method for connecting optical fibers involves a butt-joint connection. Any butt-joint requires three fundamental operations: fiber end preparation, fiber alignment to micron precision and alignment retention. Demountable connections retain alignment mechanically while permanent connections retain alignment through melting and fusing of the fiber ends in a fusion-splicing machine. In any fiber joint, the fiber ends must be prepared smooth and perpendicular to the fiber axis (as discussed in Chapter 5). The next step of aligning the fiber ends (to be jointed) is very crucial because any kind of misalignment would lead to a transmission loss. Loss at a fiber splice could originate from either or a combination of the following possible misalignments (see Fig. 10.1):

1. Transverse offset between the fiber ends
2. Angular offset between the fiber ends
3. Longitudinal end-separation between the fiber end faces.

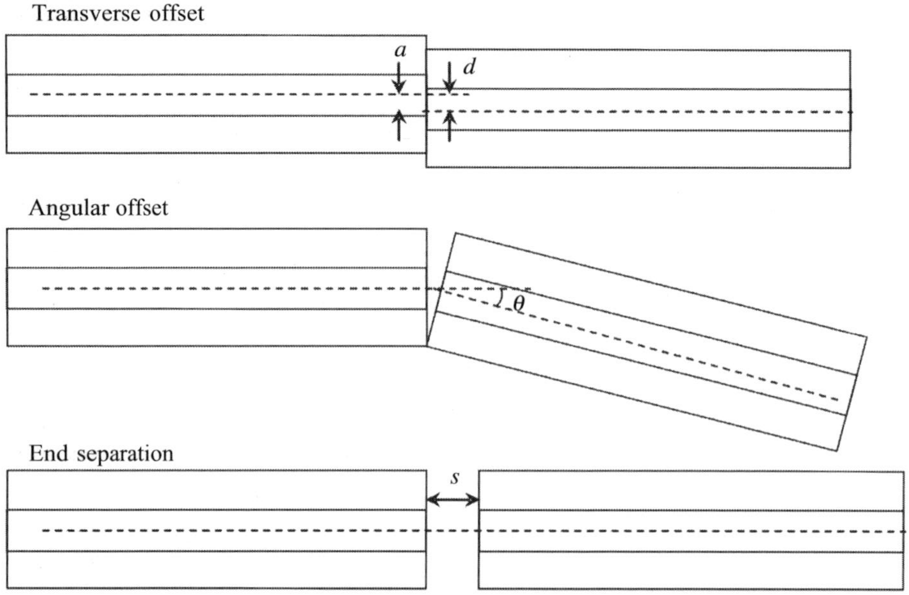

Fig. 10.1: Three different types of offsets between the axes of the fibers at a splice-joint, which may lead to a transmission loss.

These fall under the category of extrinsic losses. In addition, either or a combination of the following may also result in a joint loss and these are the sources of intrinsic loss:

1. Reflection loss at the fiber-air interfaces
2. Difference in characteristic parameters, e.g. refractive indices, index exponent, core dimensions, etc. between the two fibers
3. Deformation of the end surfaces of fibers.

Major contributors to loss, however, at a splice are transverse offset, mismatches in core radii, and the relative core-cladding index difference Δ (and hence NA). This chapter introduces Gaussian power distribution model for experimental estimation of splice losses.

10.2 Aim

To measure the power loss at a splice between two multimode fibers and study the variation of splice loss with transverse, longitudinal and angular offsets.

10.3 Apparatus

Tungsten halogen lamp/He-Ne laser, assorted microscope objectives (20X, 40X), four XYZ-translation stages, four V-groove fiber mounts, a rotation stage, optical power meter, two pieces of multimode fiber.

10.4 Theory

An accurate model of splice loss is extremely difficult to construct. Losses at a fiber splice depend on various factors like mode power distributions, attenuation, and mode coupling characteristics of the fibers. These characteristics are difficult to measure experimentally and hence several approximate models have evolved in the literature to estimate splice losses in multimode fibers. Most successful attempt in this direction has been the phenomenological model of a Gaussian power distribution [1]. "This model gives a very good agreement with measured splice losses both for individual parameter mismatches and for combinations of both intrinsic and extrinsic mismatches that are extremely difficult to calculate in any other way" [2]. In practice for a splice between two identical fibers, the following empirical but approximate expression for splice loss (in dB) due to transverse offset has been found to be adequate:

$$\Gamma_d = -0.0001 + 0.6688\left(\frac{d}{a}\right) + 4.136\left(\frac{d}{a}\right)^2 + 0.5\left(\frac{d}{a}\right)^3 \tag{10.1}$$

where d is the transverse effect (i.e. separation between the fiber axes) and a is the core radius of the fiber. It matches splice loss according to Gaussian model within 0.01 dB up to $\frac{d}{a} = 0.8$.

It has been observed that major contributors to splice loss namely, transverse offset, core radii and Δ mismatches between the transmitting and receiving fibers do not combine linearly. However, the following approximate relation may be used to model splice loss due to simultaneous presence of all the three factors [2]:

$$\Gamma = 2.0417 - 0.85k^2 - 1.035A + 5.4986\left(\frac{d}{a}\right)^2 \tag{10.2}$$

where $A = \frac{\Delta_R}{\Delta_T}$ and $k = \frac{a_R}{a_T}$; subscripts R and T refer to the receiving and the transmitting fibers. In most cases, the ratios A and k would be unity because the two fibers would be identical.

For angular offsets, transmission loss (in dB) may be approximately estimated from the following expression valid for step-index fibers [3, 4]:

$$\Gamma_\theta = 10 \log_{10}\eta \tag{10.3}$$

where

$$\eta \approx \frac{16\left(\frac{n_1}{n_{cl}}\right)^2}{\left[1+\left(\frac{n_1}{n_{cl}}\right)\right]^4}\left[1-\frac{n\theta}{\pi n_1(2\Delta)^{\frac{1}{2}}}\right] \tag{10.4}$$

n_1, n_{cl} are the core and cladding refractive indices, respectively, n is the refractive index of the medium between the two fibers and Δ is the relative core-cladding index difference.

If the end-faces of two step-index fibers to be jointed are longitudinally separated by an amount s, transmission loss (in dB) can be approximately expressed as [5, 6]

$$\Gamma_s \approx 10\log_{10}\left[1-\frac{s}{a}\frac{2}{\pi(NA)^2}\left\{\sin^{-1}(NA)-NA\sqrt{1-(NA)^2}\right\}\right] \quad \text{for } \frac{s}{a} << 1 \tag{10.5}$$

For parabolically graded-index fibers, corresponding expression is given by [5, 6]

$$\Gamma_s \approx 10\log_{10}\left[1-\frac{s}{2a}(NA)\right] \quad \text{for } \frac{s}{a} << 1,\ NA << 1 \tag{10.6}$$

We may note here that all the formulae [Eqs. (10.1)–(10.6)] are based on certain assumptions/models and care should be exercised while attempting any comparison of measurements with these theoretical expressions. *Fusion splicing* is the most widely used technique to permanently join two fibers. It is accomplished by localized heating (through an electric arc) of the interface between two butted and pre-aligned fiber ends. A high-voltage electric arc generated between two electrodes melts the fiber ends, and fuses them permanently as soon as the voltage is switched off. In the design of any such fiber fusion-splicing machine, it is essential to have a prior knowledge of the factors that may lead to splice loss. In order to minimize, splice loss, it is essential to understand relative affects of different local splice loss contributors. To appreciate it, the experiment discussed in the chapter may be performed.

10.5 Procedure

The experimental set-up is shown in Fig. 10.2. Light from a tungsten halogen lamp is launched into a piece of multimode fiber. Cladding mode strippers should be applied at the input and exit ends of this fiber; henceforth this fiber will be referred as transmitting fiber. A second piece of (preferably identical) multimode fiber, referred as the receiving fiber, is then brought near the output end of the transmitting fiber. Input end of the receiving fiber is mounted on a XYZ translation stage; the other end of this fiber is coupled to a pinhole photodetector that is connected to a power meter. One/two drops of cladding mode stripper is/are once again applied, as before, near the two ends of the receiving fiber. The XYZ-translation stages are manipulated to induce

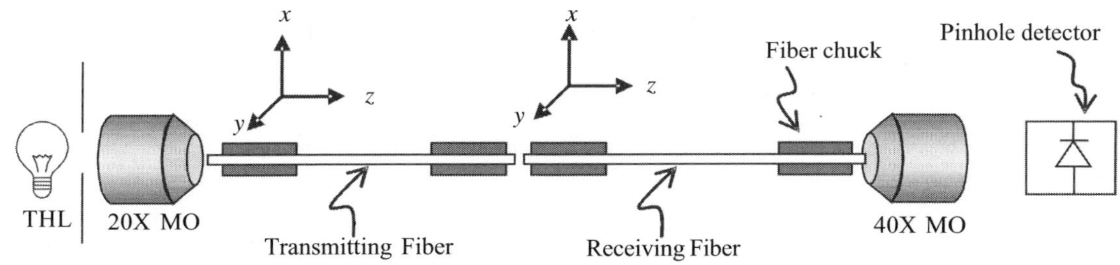

Fig. 10.2: Experimental set-up to measure fiber-to-fiber (multimode) splice loss.

any offset loss between the fibers. In actual practice for measuring transverse offset loss, it is recommended that the receiving fiber be moved away from the transmitting fiber (in the transverse direction) till the power meter reading reduces to almost zero (i.e. registers only ambient light). The receiving fiber's input end is then gradually moved into the field of view of the transmitting fiber to couple light from the transmitting fiber and the power meter reading is noted. The fiber could be moved in steps of one division of the graduated scale on the translation stage, and at each step the power meter reading is recorded. This procedure is continued till the receiving fiber once again gets out of the field of view of the transmitting fiber so that the power meter only registers background ambient light. The tabulated readings of power meter versus linear position of the receiving fiber across the joint are then used to draw a graph between these two measurement parameters. Detected power can be further converted to a dB-scale by normalizing the power meter readings for each offset with the power meter reading for zero misalignment between the fibers, which would obviously correspond to the maximum of the power meter readings. In a similar manner, the measurements could be repeated for offsets in the two other orthogonal directions. If the fiber is perfectly circularly symmetric then one of these two last measurements should yield results almost identical to the previous one. On the other hand, the third set of measurement, which would correspond to a longitudinal offset, the results would be quite different compared to transverse offset. These measurements would reveal that longitudinal offset is more tolerable than transverse offset in terms of achieving a low loss fiber joint.

Tolerance to any angular offset on a plane between the fibers can likewise be obtained by mounting the input end of the receiving fiber at the centre of a graduated turntable (i.e. rotation stage). The output end of the transmitting fiber is initially aligned in the *xy*-plane with a gap of 2~3 μm (along z) from the receiving fiber at the centre of the rotational stage. The output end of the transmitting fiber is mounted on a V-groove, which is freely suspended without contacting the rotational stage. Power coupled across the joint is then measured and recorded as a function of angular offsets. Angular offset is induced between the fiber ends by means of the rotational stage on which the receiving fiber is mounted. As before, care should be taken to use cladding mode strippers during the measurements.

10.6 Observations

A. Power loss measurement due to transverse offset:

S. No.	Transverse offset	Output power	Normalized transverse misalignment $\left[=\frac{(d_{max}-d)}{2a}\right]$	Loss (in dB) $=10\log_{10}\left(\frac{P_{max}}{P}\right)$
1. 2.				

B. Power loss measurement due to angular offset:

S. No.	Angular offset (θ)	Output power	Normalized angular misalignment $\left[=\frac{(\theta_{max}-\theta)}{\sin^{-1}(NA)}\right]$	Loss (in dB) $=10\log_{10}\left(\frac{P_{max}}{P}\right)$
1. 2.				

C. Power loss measurement due to longitudinal offset:

S. No.	Longitudinal offset	Output power	Normalized longitudinal misalignment $\left[=\frac{(s_{max}-s)}{2a}\right]$	Loss (in dB) $=10\log_{10}\left(\frac{P_{max}}{P}\right)$
1. 2.				

Note: d_{max}, θ_{max}, s_{max} correspond to the transverse, angular and longitudinal offsets for the maximum power meter reading.

10.7 Results

A typical variation of experimentally measured loss for a given multimode fiber is shown in Fig. 10.3. In this figure, filled squares show experimentally observed loss for transverse misalignment, circles for longitudinal misalignment, and triangles for angular misalignment.

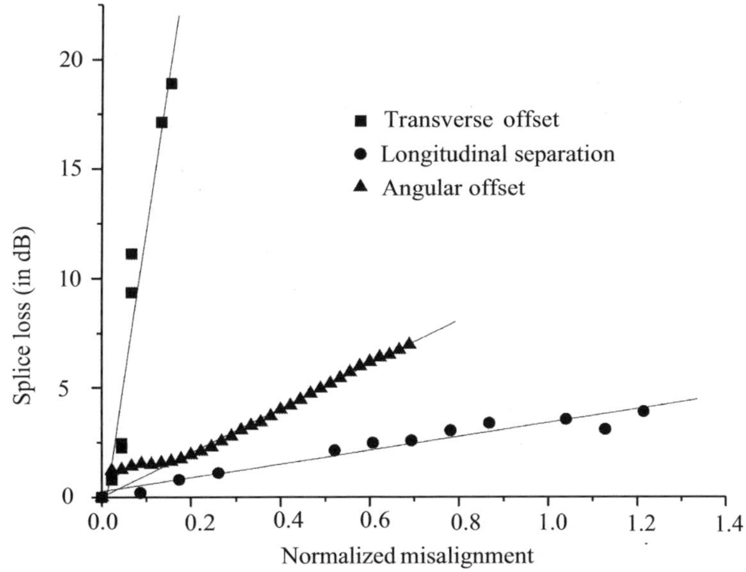

Fig. 10.3: Typical results of splice losses due to various offsets between two multimode fibers.

Loss varies almost linearly for all the three offsets as can be observed from the experimental plot. Further, the loss is most sensitive to transverse misalignment and least sensitive to longitudinal misalignment.

10.8 Comments

A comparative study of loss tolerance to various offsets at a joint between two fibers highlights that the loss, which is most critical in designing of low loss fiber splicing machines, is due to transverse misalignment. Even 10% transverse offset causes a splice loss of about 12 dB (losing 93.75% optical power!) as can be observed from Fig. 10.3. These transverse offsets are relative to the core diameter of a given fiber. Nevertheless, being the major contributor to splice loss, one is required to take extreme care to avoid any potential source of transverse misalignment while designing any splice machine.

References

1. C.M. Miller and S.C. Mettler, Bell Syst. Tech. J. vol. 57, p. 3167 (1978), and S.C. Mettler, *ibid*, vol. 58, p. 2163 (1979).
2. C.M. Miller, S.C. Mettler and F.A. White, *Optical Fiber Splices and Connectors: Theory and Methods,* Marcel Dekker Inc., New York (1986).
3. H. Tsuchiya, H. Nakagome, N. Shimizu and S. Ohara, *"Double Eccentric Connectors for Optical Fibers*", Applied Optics, vol. 16, p. 1323 (1977).
4. J.M. Senior, *Optical Fiber Communication: Principles and Practice,* Prentice-Hall International, London (1985).
5. W. Van Etten and J. Van Der Plaats, *Fundamentals of Optical Fiber Communications*, Prentice Hall International, Hempstead (1991).
6. W. Van Etten, W. Lambo and P. Simons, "Loss in multimode fiber connections with a gap", Applied Optics, vol. 24, p. 448 (1985).

11

Microbending Loss and Application in Sensing

B.D. Gupta and Anurag Sharma

11.1 Introduction

In contrast to macrobend loss due to a constant fiber curvature, if the fiber lay is made to go through a continuous succession of very small bends, it may exhibit a significant rise in a fiber's transmission loss known as microbending loss. Physically loss arises due to redistribution of optical power amongst the guided modes and/or transfer of optical power from some of the higher order guided modes to the radiation modes. The main sources of microbending could be periodic micro-deformations/ripple in an otherwise straight fiber that may come from improper spooling/cabling of the fiber or the fluctuation of the core diameter. However, the fiber manufacturing and cabling process is so advanced today that the loss incurred by microbending is not of any real concern. A careful handling of an optical fiber by the users further makes this loss negligible. Nevertheless, this mechanism can be exploited for realizing an optical fiber sensor. To the majority of students "fiber optics" is synonymous with optical telecommunication. But a great deal of technological advances in fiber optics has made remarkable impact on sensing applications as well, leading to a tremendous progress in exploitation of the fiber as a versatile sensing medium. Today almost all the physical/chemical parameters of interest, e.g. pressure, temperature, liquid level, refractive index, concentration, pH, antibodies, electric/magnetic field, displacement, rotation, etc. can be measured employing optical fiber as sensors. The greatest advantage of an optical fiber sensor over the traditional sensors lies in its capability of remote/distributed sensing and multiplexing of the sensed signals, immunity to electromagnetic interference and high sensitivity. Broadly, optical fiber sensors are classified either as *intrinsic* or *extrinsic* sensors depending on the employed sensing mechanism. In an *intrinsic* sensor, one or more of the optical properties of the fiber-guided light, such as intensity (amplitude), phase, or state of polarization is modulated by the measurand, which is then detected at the output end of the fiber through demodulation. In contrast in an *extrinsic* sensor, the fiber serves only as a conduit to carry light signal – both for the incident light to the sensor head/probe and the sensed modulated light to the detection system via the same or a second fiber. This chapter introduces the readers to the concept of a microbend-based optical fiber sensor. The experimental technique to realize fiber optic sensor discussed here employs intensity modulation scheme, which is easier to realize and demonstrates some of the salient features of the microbend effects.

11.2 Aim

To study a simple intensity modulated fiber optic pressure sensor based on microbending loss in a multimode fiber.

11.3 Apparatus

Laser diode/He-Ne laser, photodetector with multimeter, photodetector holder, microbend deformer, microscope objective (20X), microscope objective holder, two XYZ-translational stages, two fiber chucks, two post bases and posts, weight box, about 2 m length of a multimode fiber, razor blade, fiber cutter, index matching liquid.

11.4 Theory

This experiment describes a simple intrinsic fiber optic sensor, which is based on the intensity modulation of light transmitted through a fiber by inducing microbends along the fiber using periodic deformer element. When a portion of a fiber-lay is sandwiched between two deformers and pressure is applied to one of these deformers, fiber undergoes periodic deformation in the form of microbends (see Fig. 11.1). The resultant mechanical deformation of the optical fiber,

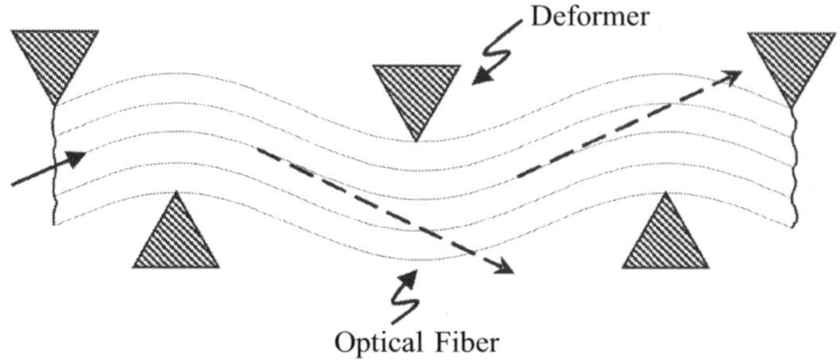

Fig. 11.1: Principle of microbend-induced loss in an optical fiber.

perpendicular to its axis, causes higher-order guided modes to radiate out of the fiber's core through the core-cladding interface as shown in Fig. 11.1. This results in the drop of transmitted light intensity through the fiber with increasing deformations.

Figure 11.2 shows the basic geometry of the microbend element about one (upper) deformation point. As the pressure is applied (from the top) on the deformer, microbending occurs along the length of the fiber, and this bending results in the loss of guided power by radiation at the bend. The loss in amplitude for a bent fiber is given by [3]

$$loss = C\left(\frac{a}{R}\right)^2 \tag{11.1}$$

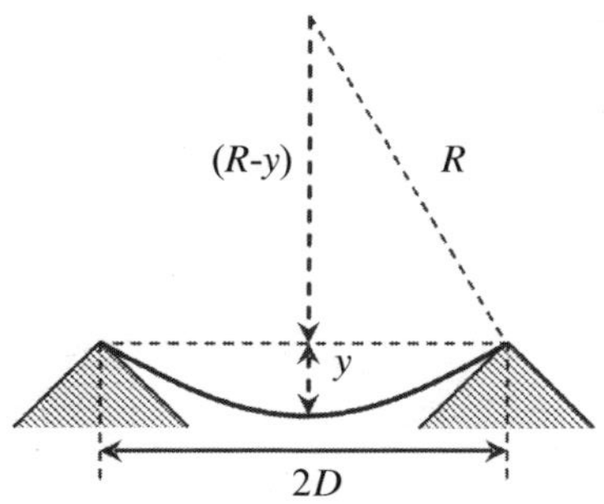

Fig. 11.2: Geometry of the microbend.

where R is the radius of curvature of the bend, a represents the fiber core radius and C is a constant. Thus, for a given fiber the pressure applied can be related to the bend radius, which is given by (see Fig. 11.2)

$$R = \frac{(y^2 + D^2)}{2y} \tag{11.2}$$

where y is the displacement of the deformer element and $2D$ is the distance between the contact points of the deformer element, which is equal to the pitch of the element. Thus, the transmittance T through the fiber is

$$\begin{aligned} T &= 1 - loss \\ &= 1 - \frac{Ca^2}{\left(\dfrac{y^2 + D^2}{2y}\right)^2} \\ &= 1 - C'\left(\frac{q}{1+q^2}\right)^2 \end{aligned} \tag{11.3}$$

where
$$C' = \frac{4Ca^2}{D^2} \tag{11.4}$$

and
$$q = \frac{y}{D} \tag{11.5}$$

The applied force and hence the pressure is proportional to the displacement y. Therefore in terms of pressure, we have

$$q = \frac{PA}{kD}. \tag{11.6}$$

where P is the pressure, A is the surface area of the deformer and k is a constant.

11.5 Procedure

Figure 11.3 shows the basic configuration of a fiber optic microbend displacement sensor. To carry out the experiment, following procedures have to be followed step by step:

1. Mount the laser source on the aligner and adjust with the help of the aligning screws.
2. Prepare the fiber ends so that it has well-cleaved ends, and clamp over the fiber chucks with magnets. One end of the fiber is fixed on the XYZ-translational stage and other on a post base.
3. Light is launched from the laser diode using a 20X-microscope objective into the fiber (as discussed in Chapter 5). The output end of the fiber is coupled to the photodetector, which is connected to a multimeter to measure the amount of light transmitted through the fiber.
4. Microbend deformer is introduced somewhere along the length of the fiber as shown in the experimental set-up. By gently pressing on the deformer element, observe drop in the transmitted light intensity through the fiber.

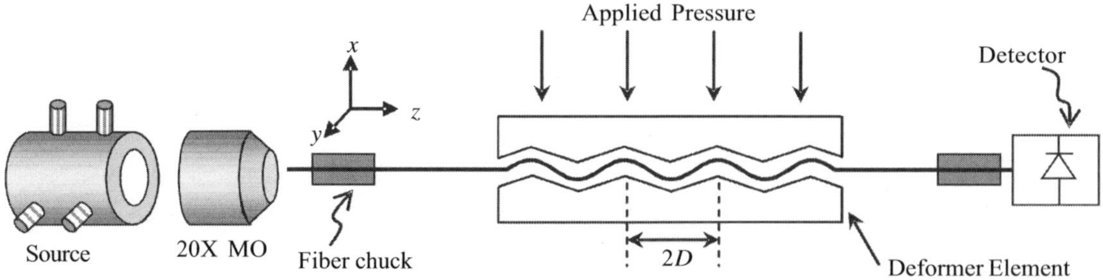

Fig. 11.3: Experimental set-up for a simple intensity modulated fiber optic pressure sensor using a microbent multimode fiber.

5. Apply different pressures by placing suitable weights on the deformer. Note down the multimeter reading corresponding to the output of the detector each time.

6. Plot the measured relative output powers as a function of the applied weights.
7. Best fit the theoretical expression for transmittance (T) (Eq. 11.3) vs. weight to the experimentally observed data by adjusting C' as the fitting parameter.

11.6 Observations

S. No.	Weight placed on the deformer	Output power	Normalized output power
1.			
2.			
3.			
.			
.			

11.7 Result

Figure 11.4 shows a graph of the measured light transmittance through a microbend-modulated fiber optic sensor versus the applied weight. First one (11.4 a) shows the data observed for the jacketed multimode fiber (core diameter 49 μm, NA 0.19), whereas the second one (11.4 b) shows the data observed for the same multimode fiber but with the jacket removed. Pitch (D) of the deformer element 1 is 1 mm and of deformer element 2 is 2.15 mm in both the cases. In this figure, filled squares and circles represent experimentally observed values and the solid curves are the least square fits to the data using expression for the transmittance (T) (Eq. 11.3) with the adjustable parameter C'. As can be observed, the transmittance varies almost parabolically with the applied pressure.

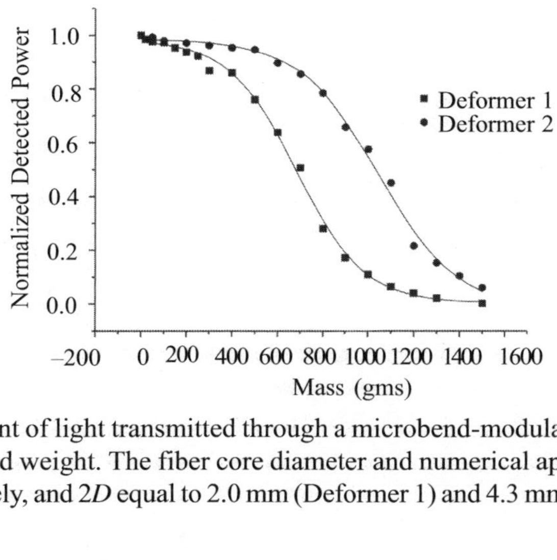

Fig. 11.4: (a) Amount of light transmitted through a microbend-modulated fiber optic sensor versus the applied weight. The fiber core diameter and numerical aperture were 49 μm and 0.19, respectively, and $2D$ equal to 2.0 mm (Deformer 1) and 4.3 mm (Deformer 2).

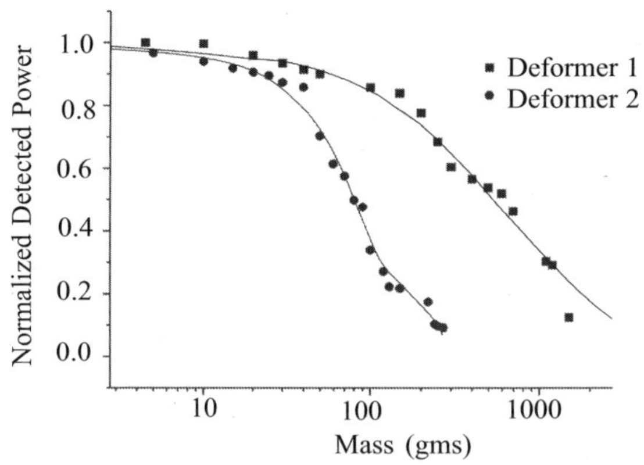

Fig. 11.4: (b) Normalized light transmission through a microbend-modulated fiber optic sensor versus the applied weight. The fiber used in this experiment was without the primary jacket, having the same core diameter and numerical aperture as in (a), with 2D again equal to 2.0 mm and 4.3 mm.

11.8 Comments

In multimode fibers, it is usually the highest order modes that get coupled to the radiation modes, and suffer the microbend loss. As these higher order modes carry only a small fraction of the total power, step-index multimode fibers are less sensitive to microbending as compared to single-mode fibers. Thus, while realizing a pressure sensor, microbending sensitivity of the sensing fiber must be maximized. One of the most important parameters in determining the microbending sensitivity is the periodicity of the fiber deformation. To understand this, one has to consider the mode coupling effects in fibers. According to the couple mode theory, introduction of periodic microbends along the fiber axis leads to the light power coupling between modes with longitudinal propagation constant β and β' satisfying [4]

$$\beta - \beta' = \pm\frac{2\pi}{\Lambda} \tag{11.7}$$

where Λ is the period of the mechanical perturbation. For the fiber of given refractive index profile, the detailed mathematical analysis predicts [5] that the higher order modes can be coupled with smaller periods Λ, whereas lower order modes are coupled with larger periods. For small changes in the fiber deformation where neighbouring mode coupling is believed to be valid, only adjacent modes are coupled. In this case, microbend-induced loss is due to the coupling of highest order guided mode to a radiation mode, as mentioned above. Such coupling can be realized with periodicities predicted by Eq. (11.7) [6], which significantly enhances microbending loss and hence the sensitivity.

In designing the microbend sensor it is important to consider the macrobending effect, i.e., light power loss caused by fiber deformations of the order of centimeters. Even though such macrobends cannot introduce mode coupling (which requires small bends), they extend the field of the propagating modes into the coating, which in turn can introduce losses. In particular, macrobending removes the preferred higher order core modes, which are responsible for the microbending sensitivity. Therefore macrobending can significantly decrease the microbending effect and must be avoided by keeping the sensing fiber straight.

The length of the sensing fiber is an important design parameter, which influences the sensitivity of the sensor. It has been found that the microbending sensitivity varies approximately linearly with the length of the sensing fiber [6]. For obtaining the best sensor response, following additional points must be ensured, at the time of conducting the experiment:

1. The fiber end faces must be of good quality.
2. Cladding-mode strippers must be used.
3. The two plates of the deformer must be accurately aligned.
4. The sensor response is critically dependent on the excitation conditions. To ensure that all the guided modes are excited in the multimode fiber, an overfill launch is applied at the input end of the fiber i.e. one uses a microscope objective of NA higher than that of the fiber for launching light into the fiber.

Finally, we may mention that although the microbend sensitivity of a fiber without the primary jacket is much higher than that with the jacket, the fiber is prone to breakage in the former case due to intrinsic brittleness associated with glass. Hence, usually microbend sensors employ fibers by retaining the primary jacket.

References

1. B. Culshaw, *Fiber Optic Sensors and Signal Processing*, Peter Peregrunus, Stevenage (1984).
2. B.P. Pal (Ed.), *Fundamentals of Fiber Optics in Telecommunication and Sensor Systems*, Wiley Eastern Ltd., New Delhi (1992).
3. C.K. Kao, *Optical Fiber Systems: Technology, Design and Applications*, McGraw-Hill, New York (1982).
4. A.K. Ghatak and K. Thyagarajan, *Introduction to Fiber Optics*, Cambridge University Press, Cambridge (1998).
5. C.N. Kurtz and W. Streifer, *IEEE Trans. Microwave Theory Tech. MTT*, vol. 17, p. 250 (1969).
6. N. Lagakos, J.H. Cole, and J.A. Bucaro, "Microbend fiber optic sensor", *Applied Optics*, vol. 26, p. 2171 (1987).

12

Characterization of Directional Coupler

M.R. Shenoy and Bishnu P. Pal

12.1 Introduction

One of the most important advantages and unique feature of optical fibers is their ability to simultaneously transport multiple signals over a single fiber in the 1300–1600 nm spectral bands and then redistribute the signals across multiple fiber terminals, wherefrom the signals are directed to specific target receivers. This process of simultaneously transporting a number of signal carrying wavelengths along a single fiber is known as *wavelength-division-multiplexing* (WDM) and corresponding optical communication systems are called WDM systems. To make such multi-terminal WDM systems functional, precisely tuned passive components are needed for combining and/or splitting signals over the fiber network. Fiber optic couplers are one of the key building blocks for such WDM fiber optic communication systems. These are the devices used for splitting or combining the signals in optical domain and generally allow light to travel through in one target direction, hence termed as *directional coupler*. Couplers could have any desired number of input and output ports (M × N coupler). In the most simple and fundamental configuration, a coupler has two input and two output ports (2 × 2 coupler) and it is used as a power splitter or combiner for light, e.g. for a He-Ne laser light or light from a laser diode. While deploying this fundamental WDM component in an optical fiber network, one needs to know its performance characteristics such as the ratios in which the power carried by the signal would be split to name a few. Thus, characterization of a coupler becomes essential while designing a communication link. This chapter introduces the basic theory of a directional coupler and the experimental technique to evaluate its various important characteristics.

12.2 Aim

To determine the characteristic parameters of the given biconical fused fiber power splitter.

12.3 Apparatus

Laser diode/He-Ne Laser, laser diode aligner, microscope objective (20X), microscope objective holder, XYZ-translational stage, photodetector connected to a multimeter, photodetector holder, two fiber chucks, two post bases and 3 posts, directional coupler, razor blade, fiber cutter, index matching liquid.

12.4 Theory

The most widely used fiber coupler is a four port device with 2 input ports and 2 output ports (Fig. 12.1), formed by fusing 2 single-mode fibers over a selected length with an oxy-butane flame or a resistance heated furnace and is often referred to as a *single-mode fiber directional coupler*. During its fabrication, two fibers are twisted around by a certain number of turns and are held in intimate contact over a short unjacketed length by means of two motorized linear translational stages. This twisted portion of the fibers is heated and slowly pulled apart simultaneously to induce a certain amount of tension. This leads to elongation of the fused region and a consequent decrease in its cross-section, so that a fused *down-taper* is formed at one end and an identical *up-taper* at the other end (see Fig. 12.2). Since the input and the output tapers are identical, either may be used as "input end". In view of this, a fused fiber directional

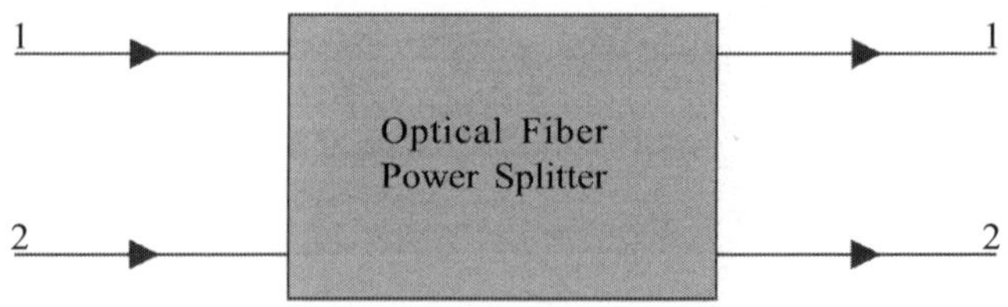

Fig. 12.1: Optical fiber directional coupler with M (here $M = 2$) number of input ports and N (here $N = 2$) number of output ports.

coupler is also often referred to as Fused Biconical Taper (FBT) fiber coupler. In the fused and tapered region, the two fibers optically interact through leakage of the evanescent field of the mode associated with the throughput fiber into the coupled fiber since the claddings in that region become extremely thin due to heating and pulling. Depending on the length of the taper, part/whole of its input light gets coupled to the second fiber. For a relatively longer taper length, the coupled light may switch back to the throughput fiber. Thus, in a fiber directional coupler depending on the length of the taper (which is controlled during fabrication), one can arbitrarily distribute the optical power launched at one of its input ports between the two output ports. In the following section, we define various basic characteristics of a fused fiber directional coupler.

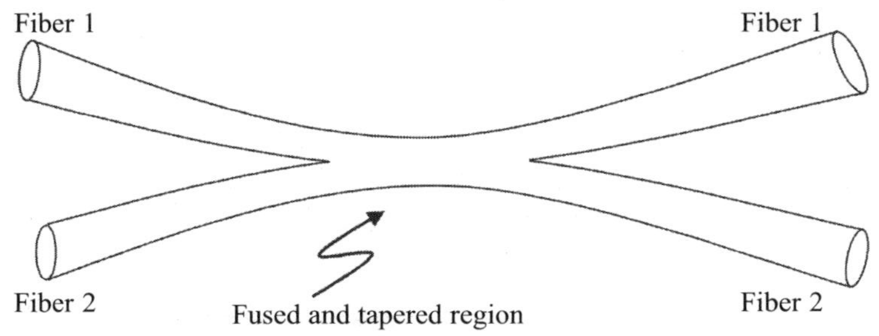

Fig. 12.2: Optical fiber fused biconical tapered (FBT) coupler.

12.4.1 Characteristic Parameters

Figure 12.3 shows the schematic of a single-mode fiber directional coupler. Port 1 and Port 4 are the input ports, and Port 2 and Port 3 form the output ports of the device. P_1 and P_4 represent the input powers at Ports 1 and 4, and P_2 and P_3 represent the output powers at Ports 2 and 3, respectively. Light is normally launched into the device through Port 1, due to which the portion AB is called the "input arm" of the coupler. From this analogy, other three port sections of the coupler have been named "throughput arm", "coupled arm" and "reflected arm" as shown in Fig. 12.3. This is the most common configuration of a directional coupler. Using the above notations, we define the characteristic parameters/specifications of a coupler as follows:

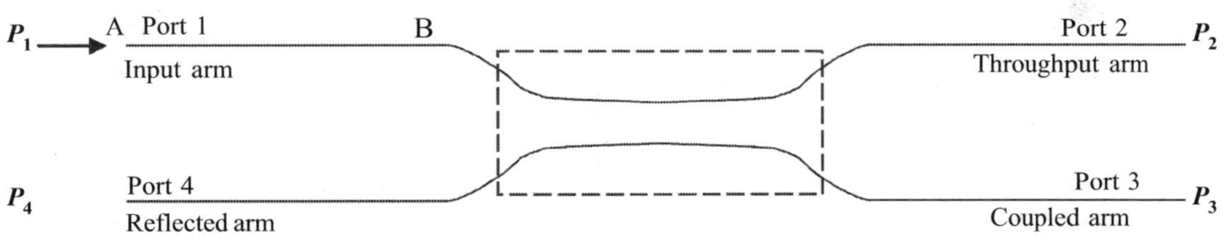

Fig. 12.3: Schematic of a single-mode fiber directional coupler.

1. **Excess Loss (*EL*):** Expressed in dB, it is the ratio of the total output power available at the two output ports to the launched power P_1. Thus,

$$EL\ (\text{dB}) = -10\log\left(\frac{P_2+P_3}{P_1}\right) \tag{12.1}$$

Commercially available couplers typically exhibit excess loss in the range 0.1–0.4 dB; couplers with even lower excess loss figures are also available. For more than two output ports

$$EL\ (\text{dB}) = -10\log\left(\sum_{r=1}^{n}\frac{P_r}{P_1}\right)$$

EL is a figure of merit to represent the overall efficiency of the coupler.

2. **Coupling Ratio (*CR*):** It is the ratio of the power in the coupled arm to the total power available at all the output ports, multiplied by 100 to express as a percentage

$$\text{Coupling ratio (\%)} = \frac{P_3}{P_2 + P_3}\times 100 \tag{12.2}$$

It describes the splitting of launched light between the output ports. For symmetric fused fiber couplers, formed by the lateral fusion of two identical single-mode fiber couplers, typical values of the coupling ratio may vary from 1% to 99%.

3. **Splitting Ratio (*SR*):** It is the ratio of the powers in the coupled arm to that in the throughput arm. A convenient means to express the splitting ratio is given by the ratio of fractional output power in the coupled arm and the fractional output power in the throughput arm, multiplied by 100.

$$\text{Splitting Ratio (\%)} = \left(\frac{P_3}{P_2 + P_3}\right)\times 100 : \left(\frac{P_2}{P_2 + P_3}\right)\times 100 \tag{12.3}$$

Thus, couplers are specified by $\frac{50}{50}, \frac{80}{20}, \frac{30}{70} \ldots$ etc. in terms of the splitting ratio. Thus, a $\frac{80}{20}$ coupler would imply that 80% of the total output power appears at Port 3 in the coupled arm, and 20% appears at Port 2 in the throughput arm. From the point of view of each output port, the power appearing at one output port is a loss to the other output port. Thus, one can define a splitting loss in dB as

$$\text{Splitting loss (dB)} = -10\log\left(\frac{P_2}{P_2 + P_3}\right) \tag{12.4}$$

associated with the throughput arm, and

$$\text{Splitting loss (dB)} = -10\log\left(\frac{P_3}{P_2 + P_3}\right) \tag{12.5}$$

associated with the coupled arm.

4. **Insertion Loss (*IL*):** It is defined as the fractional power transferred from an input port to an output port along a particular path; in the unit of decibels it is given as

$$\text{Insertion loss (dB)} = -10\log\left(\frac{P_2}{P_1}\right) \tag{12.6}$$

for the throughput arm (i.e. along the path Port 1 to Port 2), and

$$\text{Insertion loss (dB)} = -10\log\left(\frac{P_3}{P_1}\right) \tag{12.7}$$

for the coupled arm (i.e. along the path Port 1 to Port 3). Note that for each path,

$$\text{Insertion loss (dB)} = \text{Excess loss (dB)} + \text{Splitting loss (dB)} \tag{12.8}$$

The insertion loss of a directional coupler may vary from a fraction of a decibel to several decibels for any particular path depending on the splitting ratio. It represents the total loss suffered by the light while propagating from any specific input port, through the coupler, to a specific output port. It is a key performance parameter of interest to a system designer, who is concerned with calculations of power budget. It represents the *throughput loss* introduced by the component to the system as a whole.

5. **Directivity:** It is the *ratio* of the power appearing in the reflected arm (at Port 4) to the power coupled into the input arm (at Port 1); it is also referred to as *directional isolation*. In decibels, it is given as

$$\text{Directivity (dB)} = -10\log\left(\frac{P_4}{P_1}\right) \tag{12.9}$$

Since the power appearing at Port 4 is due to the light energy propagating in the "reverse" direction (as compared to the direction of the input power at Port 1), directivity forms a measure of the directional isolation. Ideally, one would like all the input energy to appear at the output Ports 2 and 3 with a certain splitting ratio. However, the unavoidable scattering and radiation losses in the coupling region lead to a finite excess loss and directivity of the device. The power appearing at Port 4 is actually a fraction of the scattered power that is guided along the reflected arm in the reverse direction, and usually, it is very small in magnitude. Thus, commercially available couplers have directivity figures < -40 dB. Directivity is a measure of how well the coupler transmits light to the targeted port(s), and quantifies the light reflections generated inside the coupler.

It may be mentioned that the parameter *directivity* is also known as "*near-end cross talk*" since Port 4 forms the nearer end with respect to Port 1. In general, *cross talk* refers to the ratio of the unwanted signal to the wanted signal at a particular port. Again, ideally one would like all the input power to flow in the forward direction, i.e., towards

Port 2 and Port 3, and therefore the backscattered/reflected power appearing at Port 4 forms an unwanted component, hence the name *cross talk*. It is defined as in Eq. (12.9), but without the minus sign.

Since the strength of coupling, or the amount of energy coupled from one fiber to the other is a strong function of the wavelength of light, the characteristic parameters of a directional coupler are wavelength dependent, and hence need to be specified at the operating wavelength.

Before we discuss the experimental procedure, we consider the following example to illustrate the determination of various characteristics of a coupler at a given wavelength:

Example: Let's assume that 100 μW of optical power is launched into the Port 1 of a particular single-mode fiber directional coupler. The powers received at the throughput port, the coupled port and the reflected port are, respectively, 56.5 μW, 38.3 μW and 1.7 nW. The excess loss, the splitting loss, the splitting ratio, the insertion loss, and the directivity of the coupler can be determined through the following:

(a) $$\text{Excess loss} = -10\log\left(\frac{56.5\ \mu\text{W} + 38.3\ \mu\text{W}}{100\ \mu\text{W}}\right) = 0.23\ \text{dB}.$$

(b) $$\text{Coupling ratio} = \frac{38.3\ \mu\text{W}}{56.5\ \mu\text{W} + 38.3\ \mu\text{W}} \times 100 = 40.4\%.$$

(c) $$\text{Splitting ratio} = \frac{\dfrac{38.3\ \mu\text{W}}{56.5\ \mu\text{W} + 38.3\ \mu\text{W}} \times 100}{\dfrac{56.5\ \mu\text{W}}{56.5\ \mu\text{W} + 38.3\ \mu\text{W}} \times 100} = \frac{40.4}{59.6} \approx \frac{40}{60}.$$

(d) $$\text{Splitting loss} = -10\log\left(\frac{56.5\ \mu\text{W}}{56.5\ \mu\text{W} + 38.3\ \mu\text{W}}\right) = 2.25\ \text{dB. (throughput port)}$$

$$\text{Splitting loss} = -10\log\left(\frac{38.3\ \mu\text{W}}{56.5\ \mu\text{W} + 38.3\ \mu\text{W}}\right) = 3.94\ \text{dB. (coupled port)}$$

(e) $$\text{Insertion loss} = -10\log\frac{56.5\ \mu\text{W}}{100\ \mu\text{W}} = 2.48\ \text{dB. (input} \rightarrow \text{throughput port)}$$

$$\text{Insertion loss} = -10\log\frac{38.3\ \mu\text{W}}{100\ \mu\text{W}} = 4.17\ \text{dB. (input} \rightarrow \text{coupled port)}$$

Note: Insertion loss = Splitting loss + Excess loss.

(f) Directivity $= -10\log\dfrac{1.7\ \text{nW}}{100\ \mu\text{W}}$ (or directional isolation) = 47.70 dB.

(or Near-end cross talk) = –47.70 dB.

Note: In the above example, the splitting ratio of the coupler is approximately $\dfrac{40}{60}$. If the splitting ratio was $\dfrac{50}{50}$, then the splitting loss at each of the output ports would have been 3 dB, and therefore $\dfrac{50}{50}$ couplers are also referred to as 3 dB couplers.

12.5 Procedure

The characteristic parameters of a directional coupler at a given wavelength are usually determined by the cutback method. Figure 12.4 shows the experimental set-up for this method. In the following, we outline the step by step procedure to estimate some of the characteristic parameters of a 2 × 2 FBT coupler:

1. Mount the laser diode on the aligner and adjust with the help of the aligning screws to align it with the microscope objective (as discussed in Section 1.2.2 of Chapter 5).
2. All the four fiber ends of the given directional coupler are prepared so that it has well-cleaved flat ends. The cladding modes are removed by applying an index matching liquid (e.g. liquid paraffin) over a few centimeters of the bare fiber, near all the four ends (after the removal of protective plastic jacket) and then clamped over the fiber chucks with magnets. One arm of the coupler is fixed in the XYZ-translational stage.

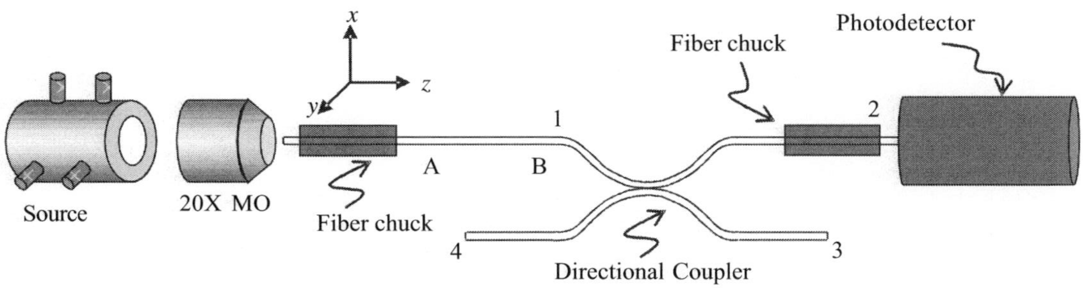

Fig. 12.4: Experimental set-up for the determination of the characteristic parameters of a fused biconical taper fiber coupler.

3. Light is launched into one arm of the given coupler from the laser diode using a 20X-microscope objective (as discussed in Chapter 5). The length of the fiber, which forms the input arm, should be at least one meter or more.

4. The output port 2 of the directional coupler is clamped over a post base and is coupled to the photodetector connected to the multimeter (which is fixed over the translational stage) to measure the amount of light transmitted through this port of the coupler. The detected power P_2 recorded by the multimeter is noted down.
5. Remove this output port of the coupler from the post base and clamp the other output port (port 3) over it. Repeat the measurement described in step 4 for this port and note down the detected power (P_3) recorded by the multimeter.
6. Repeat the Step 5 by replacing the port 3 with port 4 of the directional coupler and record the detected power (P_4).
7. Now, without disturbing the input coupling, the fiber forming the input arm is carefully cleaved (at a point B in Fig. 12.4) about 50 cm (or more) away from the input-coupling end.
8. Put the index matching liquid over a few centimeters of the bare fiber, near this new output end, which is then clamped over the fiber chucks with magnets.
9. The output power is coupled to the photodetector, which is recorded as P_1, which is nothing but the actual power coupled at the input port of the directional coupler.
10. With so recorded powers P_1, P_2, P_3 and P_4 for the same input condition, all the characteristic parameters can be obtained as illustrated through the example in Sec. 12.4.1.

12.6 Observations

S. No.	Power at throughput arm (P_2)	Power at coupled arm (P_3)	Power at reflected arm (P_4)	Power launched at input arm (P_1)
1.				
2.				
.				
.				

12.7 Comments

A major problem with the cut-back method is that it is not a nondestructive method; every time a portion of the input arm is "cut-back", and the length of the fiber would soon become too short to employ the cutback method. Therefore when the fiber coupler is procured to perform the above characterization experiment in a student-training laboratory, one must ask for an extra long length of the fiber in one or two arms of the coupler. Alternatively, one can always splice a long length of the fiber (say 20 or 30 m) to the input arm of the available couplers of standard fiber length.

References

1. A.K. Ghatak and K. Thyagarajan, *Introduction to Fiber Optics,* Cambridge University Press, Cambridge (1998).
2. B.P. Pal, Chapter *All-fiber Guided Wave Components* in "Electromagnetic Fields in Unconventional Materials and Structures", Eds. A. Lakhtakia and O.N. Singh, John Wiley & Sons, New York, 2000.

PART III

13

V-Parameter of a Single-Mode Fiber

Arun Kumar and R.K. Varshney

13.1 Introduction

Importance of *V*-parameter is emphasized in Chapter 4. Further, in Chapter 7 we have described a simple method to estimate the Gaussian mode field diameter of the fundamental mode of a single-mode fiber. In the case of a step-index single-mode fiber, a careful measurement of the far-field distribution can yield other important parameters such as the core radius, *V*-number and Δ-the relative core-cladding index difference. With a relatively low-sensitivity detector, the measured far-field intensity distribution appears like a Gaussian (see Fig. 7.4). However, the actual far-field distribution has side lobes, although very weak on the either sides of the central maximum. The angles at which the minima occur can be exploited to estimate some important parameters of the fiber.

13.2 Aim

To determine the *V*-parameter, the core radius and core-cladding dielectric constant difference of a step-index single-mode fiber.

13.3 Apparatus

He-Ne laser, two meters of single-mode (single-moded at He-Ne wavelength) fiber, 20X MO, XYZ-stacks, V-grooves, a pinhole detector connected to a power meter.

13.4 Theory

The fundamental mode field distribution of a step-index single-mode fiber is given as [1]

$$\psi(r) = \frac{AJ_0\left(\frac{Ur}{a}\right)}{J_0(U)} \quad \text{for} \quad r \le a$$

$$= \frac{AK_0\left(\frac{Wr}{a}\right)}{K_0(W)} \quad \text{for} \quad r \ge a \tag{13.1}$$

where A is a constant, and $U^2 = a^2(k_0^2 n_1^2 - \beta^2)$, $W^2 = a^2(\beta^2 - k_0^2 n_2^2)$. Here n_1 and n_2 represent the core and cladding indices, a is the core radius, $V^2 = U^2 + V^2 = k_0^2 a^2 (n_1^2 - n_2^2)$ and β is the propagation constant of the fundamental mode. It can be analytically shown that the corresponding far-field intensity pattern is given by (see e.g. [1]):

$$I(\alpha) = A^2\left[\frac{U^2 W^2}{(U^2-\alpha^2)\,(W^2+\alpha^2)}\left(J_0(\alpha) - \alpha J_1(\alpha)\frac{J_0(U)}{U\,J_1(U)}\right)\right]^2 \quad \text{for} \quad \alpha \ne U$$

$$I(\alpha) = A^2\left[\frac{U^2 W^2}{2\,V^2}\frac{1}{U\,J_1(U)}\left(J_0^2(\alpha) + J_1^2(\alpha)\right)\right]^2 \quad \text{for} \quad \alpha = U \tag{13.2}$$

where α represents the normalized radiation angle defined as

$$\alpha = k_0 a \sin\theta \tag{13.3}$$

and θ is the angle made by the direction of observation from the fiber axis in the far-field pattern. If one plots the far-field intensity distribution given by $I(\alpha)$, a typical plot appears as shown in Fig. 13.1. We assume that θ_h and θ_x respectively represent the angles in the far-field pattern at which the intensity within the main lobe is decreased to half of its maximum value and at which the first minimum occurs. It is evident from Eq. (13.2) that the corresponding values of α_h and α_x are functions of V alone. This observation can be exploited to determine the core radius and V-parameter of the fiber in the following manner [2].

Since α_h and α_x are the functions of V alone, $\frac{\sin\theta_x}{\sin\theta_h}$ i.e. $\frac{\alpha_x}{\alpha_h}$ would also be a function of V only.

Thus, if one plots α_h and $\frac{\sin\theta_x}{\sin\theta_h}$ as function of V, one would get two universal curves (1 and 2) for a step-index fiber as shown in Fig. 13.2.

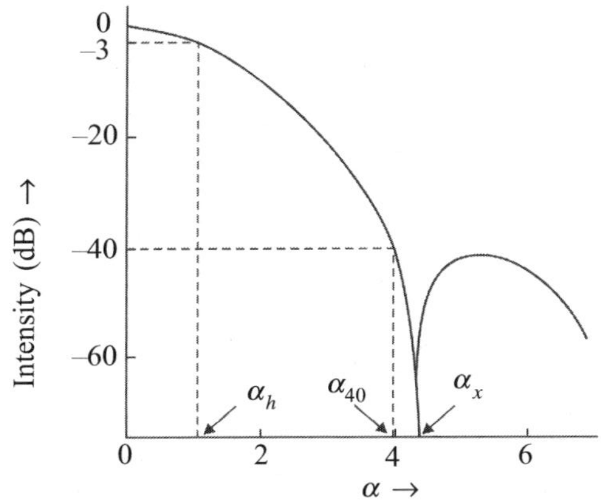

Fig. 13.1: Far-field intensity pattern for a single-mode step-index fiber with $V = 2.4$ as obtained by using Eq. (13.2).

Using these universal curves and measured values of θ_h and θ_x, one can obtain fiber parameters as discussed below:

1. Obtain the ratio $\dfrac{\sin\theta_x}{\sin\theta_h}$ experimentally. Use curve 1 of Fig. 13.2 to obtain the value of the *V*-parameter.
2. Once the *V*-parameter is known, use curve 2 of Fig. 13.2 to obtain $\alpha_h(=k_0\ a\ \sin\theta_h)$ and since $k_0 = \left(\dfrac{2\pi}{\lambda}\right)$ and θ_h are already known, calculate the core radius using Eq. (13.3).
3. Knowing *V* and *a*, the core-cladding dielectric constant difference can also be obtained using the relation

$$V^2 = k_0^2 a^2 (n_1^2 - n_2^2) \tag{13.4}$$

or

$$\Delta\varepsilon = n_1^2 - n_2^2 = \frac{V^2}{k_0^2 a^2} \tag{13.5}$$

It may be mentioned, however, that using curves 1 and 2 to obtain *V* and α_h as described above may be quite inconvenient. On the other hand, one can use the following empirical relations given by Roy et al. [3] to obtain *V* and α_h from the far-field measurement data:

$$V = 8.039 - 2.347\sigma_x + 0.3329\sigma_x^2 - 2.181\times10^{-2}\sigma_x^3 + 5.40\times10^{-4}\sigma_x^4 \quad \text{for} \quad 1.3 \le V \le 2.4 \tag{13.6}$$

and

$$\alpha_h = -0.7858 + 0.994V - 0.1155V^2 \tag{13.7}$$

where

$$\sigma_x = \frac{\sin\theta_x}{\sin\theta_h}$$

We may also add here that an accurate location of θ_x is generally not possible since the intensity in the close vicinity of θ_x is below –80 dB of the central maximum, and the measurement systems are typically sensitive upto 50 dB dynamic range. Thus, in order to locate θ_x one should measure $I(\theta)$ for a few degrees beyond the first minimum on the θ-axis into the second lobe. The intensity in the vicinity (4–5° on either side) of the minimum should then be converted into amplitude with a change of sign for the amplitude in the second lobe, and plotted as a function of θ. The value of θ_x corresponding to the zero amplitude, should be read accurately from this graph. Alternatively, one can measure the angle θ corresponding to a higher intensity level of – 40 dB and use the following empirical relation [3] instead of Eq. (13.6) to determine the *V*-parameter:

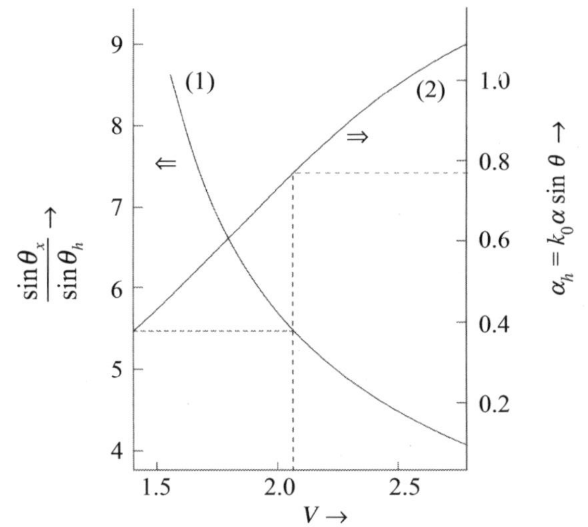

Fig. 13.2: Variation of $\frac{\sin\theta_x}{\sin\theta_h}$ (curve 1) and α_h (curve 2) with *V*-parameter for a step-index single-mode fiber.

$$V = 12.178 - 5.039\sigma_{40} + 0.9437\sigma_{40}^2 - 8.196 \times 10^{-2}\sigma_{40}^3 + 2.71 \times 10^{-3}\sigma_{40}^4$$

$$\text{for} \quad 1.3 \leq V \leq 2.4 \qquad (13.8)$$

where

$$\sigma_{40} = \frac{\sin\theta_{40}}{\sin\theta_h} \qquad (13.9)$$

Thus, by measuring θ_h and θ_x (or θ_{40}), one can obtain the *V*-parameter from the empirical relations given above, and then can calculate the core-radius *a* and the core-cladding dielectric constant difference from the following relations:

$$a = \frac{\alpha_h}{k_0 \sin\theta_h} \left(\text{or } \frac{\alpha_{40}}{k_0 \sin\theta_{40}} \right) \text{ and } \Delta\varepsilon = \frac{V^2}{k_0^2 a^2} \tag{13.10}$$

Since for practical single-mode fibers $n_1 \cong n_2 = n$ (silica), one can also calculate the relative core-cladding index difference Δ, using

$$\Delta = \frac{n_1^2 - n_2^2}{2n_1^2} \tag{13.11}$$

As mentioned in the beginning, above relations hold good for a step-index fiber only. In the case of a graded-index single-mode fiber, above technique provides the parameters of an equivalent step-index fiber known as ESI parameters.

13.5 Procedure

The experimental set-up and the procedure to plot the far-field pattern is basically the same as discussed in Chapter 7. In order to determine angles θ_h and θ_x, one has to plot the far-field pattern much more accurately with highest angular resolution, as compared to the experiment in Chapter 7. To achieve this [4],

1. An intense light source such as a laser diode (LD) should be used so that the measurements can be taken at a sufficiently large distance from the centre of the fiber output.
2. The angular sector scanned for the far-field pattern must be wide enough, around ±20° to 25° to completely include at least the main lobe of the radiation pattern.
3. The dynamic range of the detector used should be >40 dB.
4. A small area photodetector mounted on a motor-driven rotational stage should be used for the angular scan of the far-field pattern.

13.6 Observations

S. No.	θ_x	θ_h	$\frac{\sin\theta_x}{\sin\theta_h}$	V	α_h	a	Δ
1. 2.							

13.7 Results

For the given step-index single-mode fiber, V-parameter =

core radius =

core-cladding index difference =

13.8 Comments

Same as in Chapter 7. In addition, as mentioned earlier, an accurate location of θ_x is generally not possible owing to the intensity limitations. Thus, one should measure $I(\theta)$ for a few degrees beyond the first minimum on the θ-axis into the second lobe and then extrapolate in order to correctly locate θ_x.

References

1. A.K. Ghatak and K. Thyagarajan, *Optical Electronics*, Cambridge University Press, Cambridge (1989). Reprinted by Foundation Books, New Delhi (1991).
2. W.A. Gambling, D.N. Payne, H. Matsumura and R.B. Dyott, "Determination of core diameter and refractive-index difference of single-mode fibers by observation of the far-field pattern", *IEE Trans. Microwave, Optics and Acoustics*, vol. 1, p. 13 (1976).
3. S. Roy, R. Tewari and K. Thyagarajan, "Accurate empirical relations for characterizing a single-mode matched clad fiber from its far-field pattern", *J. Optical Communications*, vol. 12, p. 26 (1991).
4. Artiglia, G. Coppa, P. Di Vita, M. Potenza and A. Sharma, "Mode field diameter measurements in Single-mode optical fibers", *J. Lightwave Technol.*, vol. 7, p. 1139 (1989).

14

Loss Measurements in Single-Mode Fiber

I.C. Goyal and Sunil K. Khijwania

14.1 Introduction

When light propagates along a single-mode fiber, there is an associated loss of energy due to various mechanisms. As mentioned in Chapter 3, if P_1 and P_2 represent the input and output powers, then

$$\text{Loss of power in dB} = 10\log_{10}\frac{P_1}{P_2} \tag{14.1}$$

Let us consider an example: Input power, $P_1 = 100\ \mu$W, length of the fiber 5 km, the output power, $P_2 = 15\ \mu$W. The total attenuation in traversing the 5 km length of the fiber $= 10\log\left(\frac{100}{15}\right) \approx 8.24$ dB.

Hence, loss (per km) of the fiber $\alpha \approx 1.65$ dB/km.

Indeed the power in the fiber decreases exponentially and we may write

$$P(z) = P(0)\, e^{-\gamma z} \tag{14.2}$$

Thus, from Eq. (14.1), loss in dB $= 10\log_{10}\frac{P(0)}{P(z)} = \gamma z \times 10\log_{10}(e) \approx 4.34\ \gamma z$

Thus, a loss of 1.65 dB/km gives us

$$1.65 \approx 4.34 \times \gamma[\text{m}^{-1}] \times 1000\,[\text{m}]$$

or, $\gamma \approx 0.38 \times 10^{-3}\ \text{m}^{-1}$. In general, α (and also γ) is a function of wavelength.

14.2 Sources of Loss in silica fibers

14.2A. Absorption

Silica is the primary material component of an optical fiber used for long distance communication and pure silica has very low absorption loss in the visible and near IR wavelength region. Absorption loss in silica fibers may arise from the following factors:

1. Absorption bands due to electronic transitions occur in the UV wavelength region; the tails of these bands decay exponentially and reach into the visible region leading to absorption loss. However, besides being of no importance to optical telecommunication because these wavelengths falls below the low loss wavelength window of silica fibers, these losses are lower than the scattering loss discussed below.
2. The nearest vibration band resonance of the Si-O bond occurs in the IR wavelength region at 9.2 μm. Certain refractive index modifiers like GeO_2, P_2O_5, and B_2O_3 that are added (often referred as dopants) to silica for changing its refractive index are characterized with similar resonance at the IR wavelength region. The tails of these absorption bands in the IR may contribute in the wavelength region of interest 1.4–1.6 μm depending on the doping material.
3. Transition metals like Cu, Fe, Ni, Cr and Mn in various oxidation states, if present, in silica even at ppm level, could lead to electronic absorption loss of the propagating signal light in a fiber in or near the visible region. However, due to the very high purity materials (referred to as "opti-pure" grade) that are used as starting materials in state-of-the-art fiber fabrication methods involving vapour phase depositions, such impurities are of no concern in communication grade fibers of today.
4. Unless extreme care and certain special steps are taken during fiber fabrication, it may be difficult to completely get rid of another irritant source of impurity loss, which arises due to presence even in ppb of water in the form of radical OH^-. The OH radical has its fundamental vibration absorption band at $\lambda = 2.8\ \mu m$ with harmonics occurring at wavelengths $\cong 1.39$, 0.95 and 0.72 μm. Further subsidiary peaks could occur due to the combination overtones of the Si-OH bond. All these contribute to loss in the wavelength region of interest. Out of these most trouble shooting is the one, which appears at 1.39 μm. In low water peak fibers of current state-of-the-art, OH^- as impurity is practically absent.

14.2B. Scattering

Since the fabrication process of optical fibers involves extreme heating and rapid cooling, random minute scale density fluctuations get frozen in; this results in minute refractive index variations along the fiber length. These regions of inhomogeneities are ≤ to the wavelength (λ) of the propagating light and hence these act as source of scatterers for the propagating light. This scattered light obeys Rayleigh's law and it varies with wavelength as λ^{-4} (the law underlying the blue colour of the sky). Another kind of scattering loss could arise due to imperfections at the core-cladding interface. Due to either of these reasons, light rays are scattered in random directions so that they may enter the cladding and get lost. In terms of modes, this means some of the guided power is coupled to the radiation modes.

14.2C. Bending and microbending

In a practical fiber, due to bending and microbending (caused by pressure, coating, cabling etc.) power conversion amongst guided modes (in a multimode fiber) and to radiation modes (in single-mode as well as multimode fiber) may take place, which could result in additional loss of guided power.

The total attenuation coefficient α_t is therefore a sum of α_s (due to scattering loss), α_a (due to the absorption loss) and α_b (due to bending and microbending loss), i.e.,

$$\alpha_t = \alpha_s + \alpha_a + \alpha_b \tag{14.3}$$

It may be interesting to know that ordinary window glass has losses, which could exceed even 500 dB/km. For optical fiber communication to be practically feasible and compete with the then existing coaxial cables for carrying signals in electrical form, the initial goal was set (in late 1960's) to develop glass fibers with losses ≤20 dB/km. Astonishing progress has been made ever since, and in late 1970s, multimode fibers with loss ≤0.5 dB/km and single-mode fibers with loss ≤0.2 dB/km at λ = 1.55 μm were achieved. The theoretical limit for silica fibers is ≅0.15 dB/km at 1.55 μm.

14.3 Aim

14.3A. To determine the propagation loss in a given single-mode fiber at a given wavelength through the cut-back technique.

14.3A.1 Apparatus

Known length of single-mode fiber, two fiber chucks, one XYZ-translational stage, He-Ne laser, 20X MO, cladding mode stripper (MS), photodetector and a multimeter or a power meter.

14.3A.2 Procedure

Here we describe a simple method to determine the total attenuation coefficient in a single-mode fiber with a monochromatic source of light. Figure 14.1 shows the experimental set-up. To carry out the experiment, following procedures have to be followed step by step:

1. Adjust the incident beam at the input end "A" so that maximum output power is obtained from the other end of the fiber.
2. Measure the optical power at the output end of the fiber. Let it be $P(L)$, where L is the length of the fiber from the point "B" up to the output end.
3. Without disturbing the launching conditions cut the fiber at point "B" (a short distance after the input end) and again measure the power coming out of the fiber with the help of the photodetector. Let this power be $P(0)$.

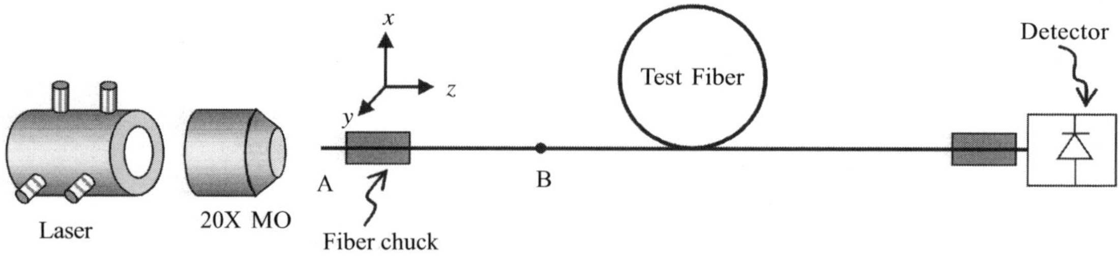

Fig. 14.1: Schematic diagram of the experimental set-up for the measurement of propagation loss in a single-mode fiber.

The attenuation coefficient, α_t of the fiber at the wavelength of the light source used will be given by

$$\alpha_t \text{ (in dB/km)} = \frac{10}{L} \log \frac{P(0)}{P(L)} \tag{14.4}$$

The method described above is known as the cut-back method, because the fiber (about 2 meters) is cut during the experiment. The cut-back method ensures that system response of the measurement set-up is removed in the measurement of loss.

14.3A.3 Observations

S. No.	Length of the fiber (L) [km]	$P(L)$ [mW]	$P(0)$ [mW]	a_t
1.				
2.				

Average α_t =

14.3A.4 Results

The propagation loss coefficient, α, in the given single-mode fiber by using the cut-back technique =

14.3B. To measure the spectral attenuation response of a given single-mode fiber.

14.3B.1 Apparatus

Known length of single-mode fiber, He-Ne laser, tungsten-halogen lamp (THL), monochromator, chopper (C), lock-in-amplifier, beam splitter, translucent screen (TS), 10X and 20X MO, cladding mode stripper, photodetectors (PD_1 and PD_2), power meter etc.

14.3B.2 Procedure

Figure14.2 depicts the experimental set-up to determine the spectral attenuation. To carry out the experiment, following procedures have to be followed step by step:

1. Get the parallel beam from the source (THL) with the help of lens "L_1", and focus it onto the end of the fiber with the help of another lens "L_2". Couple the light into the fiber as discussed in the previous chapter.
2. Arrange a monochromator and a chopper system as shown in the experimental set-up. Monochromator is used to select light of different wavelengths. Interference filters may also be used for this purpose in place of the monochromator. The chopper "C" converts the continuous beam of light into rectangular pulses of desired frequency determined by the speed of the chopper.
3. Since the intensity of light beam coming out of the monochromator is significantly less than that from a laser source, a combination of a lock-in-amplifier along with a chopper arrangement is used to increase the signal-to-noise ratio. By phase matching with the reference signal from the chopper, one only detects output power coming at the chopped frequency and thus eliminates the background, source and detector noises at frequencies other than the chopper frequency. One could also use the chopper – lock-in-amplifier combination in the previous experiment (14.3A).
4. After adjusting the incident beam at the input end so that maximum output power is obtained at any one wavelength, the following observations are taken:
   a. Monochromator is set to give light beam of different wavelengths (λ). Measure the optical power coming out from the fiber end at each wavelength. Let $P_\lambda(L)$ denote this power at wavelength λ (L is the length of the fiber from the point "B" up to the output

end). [One should be extremely careful so that the launching conditions do not change while setting the monochromator for different wavelengths.]

b. As in the previous experiment, cut the fiber at "B" without changing the launching conditions. Measure the output power again at different wavelengths. Let $P_\lambda(0)$ denote the power coming out of "B" at wavelength λ. Then

$$\alpha_t(\lambda) = \frac{10}{L}\log\frac{P(0)}{P(L)} \tag{14.5}$$

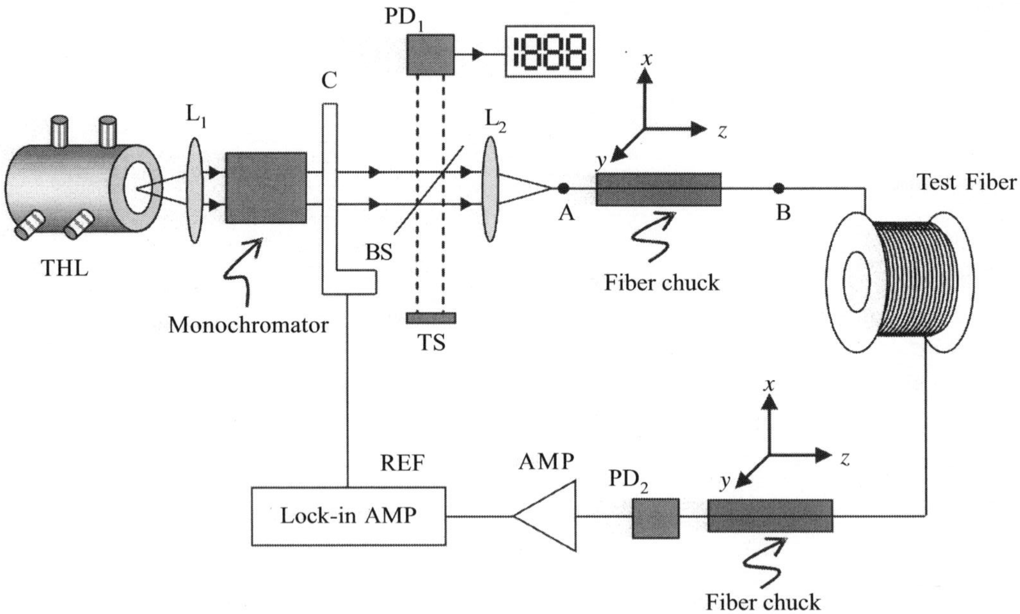

Fig. 14.2: Schematic of the experimental set-up for determination of the spectral attenuation response of a single-mode fiber.

14.3B.3 Observations

S. No.	Length of the fiber (L) [km]	Wavelength (λ) nm	$P_\lambda(L)$ [mW]	$P_\lambda(0)$ [mW]	$\alpha_t(\lambda)$
1. 2.					

14.3B.4 Results

The attenuation coefficient for the given single-mode fiber =

________ dB/km @ λ_1 = ________ nm

________ dB/km @ λ_2 = ________ nm

________ dB/km @ λ_n = ________ nm

14.4 Comments

For obtaining accurate estimation, following additional points must be ensured at the time of the experiment:

1. Beam Splitter (BS) may be used to reflect a portion of the incident light on to the photodetector (PD_1), which can be noted at each wavelength during the measurement of both $P_\lambda(L)$ and $P_\lambda(0)$. If any fluctuations in the intensity of the source occur during the experiment it can easily be accounted for.
2. To observe the location of the focused spot at the input end "A" of the test fiber, it is back illuminated by a He-Ne laser. Due to the 4% Fresnel reflection of the input beam by the input end of the test fiber (monochromator may be removed for this setting) and back illumination, one observes across the test fiber cross-section a focused spot against the laser illuminated background. This arrangement helps in adjusting the launching conditions with respect to centering the focused spot on the fiber cross-section.
3. Since the intensity of the incident light obtained by THL and monochromator arrangement is usually very small, this may be replaced by a tunable dye laser.

References

1. B.P. Pal, Chapter 4 in B.P. Pal (Ed.), *Fundamentals of Fiber Optics in Telecommunication and Sensor Systems*, New Age Publishers, New Delhi, 2002.
2. D. Marcuse, *Principles of Optical Fiber Measurements*, Academic Press, New York (1981).
3. K. Thyagarajan, B.P. Pal and Arun Kumar, "Characterization of Optical Fibers for Telecommunication and Sensors—Part I: Multimode Fibers", *Int. J. Optoelectronics*, vol. 3, p. 45 (1988).
4. R.L. Gallawa, G.E. Chamberlain, G.W. Day, D.L. Franzen and M. Young, "Measurement of multimode optical fiber attenuation", *NBS Special Publication*, Washington, vol. 2, p. 637 (1983).

15

Spectral Attenuation in a Multimode Fiber

I.C. Goyal and Sunil K. Khijwania

15.1 Introduction

In a multimode fiber, unlike the case of a single-mode fiber, it is difficult to assign loss per unit length as described by Eqs. (14.1–14.2) in the previous chapter. This is due to the fact that a large number of modes propagate in a multimode fiber and each mode is characterized by a different value of attenuation coefficient (α), termed as differential modal attenuation (DMA). As the larger percentage of light is concentrated in the outer part of the core and into the cladding for higher order modes, these modes get attenuated significantly more as compared to the lower order modes. Owing to the DMA, these higher order (outer) modes tend to die off after traversing a short length of the fiber. Another important phenomenon observed in a practical multimode fiber, is termed as "mode coupling". This results in continuous power coupling amongst guided modes as well as from guided to radiation modes. Bending, microbending, scattering and other imperfections are the primary sources for mode coupling. Therefore unlike Eq. (14.1), the total power variation will be rather complex and $P(z)$ will not necessarily decay exponentially with z in the case of multimode fiber. In order to extrapolate the value of attenuation to long fiber lengths based on measurements made on a specific short sample fiber length under laboratory environment, we need to understand the concept of the *steady state* or *equilibrium mode distribution* (SMD/EMD) with respect to multimode fibers. As the light propagates down the fiber, mode coupling takes place. The mode coupling occurs most strongly among the modes of neighbouring modal indices. Thus, modal power distribution continues to evolve with propagation till a SMD/EMD gets established, after which on a log scale the transmitted power decreases linearly with distance. This feature is best illustrated qualitatively in Fig. 15.1, where the guided power in a typical multimode fiber is plotted (on log scale) as a function of the distance of propagation. The dashed curve corresponds to the launching conditions such that more power is launched into higher order modes (overfilled i.e. overfilled N.A. launching) while dot-dashed curve corresponds to the launching conditions wherein more power is coupled into the lower order modes (under-filled launching). Under overfilled launch condition, higher order lossy modes get attenuated within a relatively short distance of propagation, after which SMD is established. For an under-filled distribution, the attenuation is less than the steady state value because preferentially lower order modes are excited, which are relatively less lossy. Thus, SMD/EMD is reached after a relatively longer length of the fiber. A balance is reached between mode coupling and DMA after a distance $z > z_c$, and the relative power distribution amongst various guided modes becomes constant. Both the curves would merge after z_c, beyond which the power decays exponentially thereafter, which can be described by an equation of the form [Eq. 14.2]

$$P(z) = P(0)\, e^{-\gamma z} \tag{15.1}$$

If transmitted power described by Eq. (15.1) (on a log scale) is plotted versus the fiber length it would be a straight line. This represents equilibrium mode distribution (EMD) or steady state mode distribution (SMD). The significance of EMD/SMD lies in the fact that the total power now decays in accordance with Eq. 15.1, and a unique loss coefficient can now be defined (see Fig. 15.1) after $z > z_c$. The *steady state* mode distribution depends primarily on the dimensions of

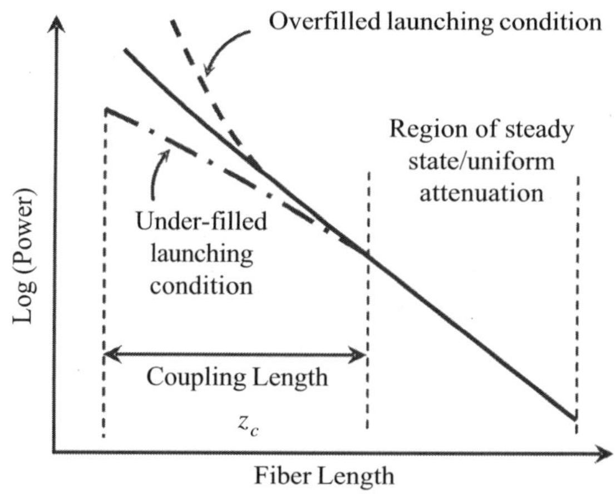

Fig. 15.1: Guided power (in Log scale) versus the distance of propagation in a multimode fiber for different input launching conditions.

the fiber, the fiber index distribution, the nature and degree of surface roughness or other imperfections, and the number and radii of bends in the fiber. Variations in these parameters, of course, will generate a variation in z_c. In high quality fibers, free from imperfections such as diameter fluctuations and microbending, SMD/EMD may be attained only after a few kilometers.

15.2 Aim

To measure the spectral attenuation response of a given multimode fiber.

15.3 Apparatus

Known length of multimode fiber, He-Ne laser, THL, interference filter (IF), chopper (C), lock-in-amplifier, beam splitter, translucent screen (TS), cladding mode stripper, photodetectors, power meter, aperture, etc.

15.4 Theory

The most significant attenuation in the case of multimode fiber is the *steady state* attenuation. Once the steady state condition is achieved, cut-back method for loss measurement described in the previous chapter where a small piece of the fiber is cut from near the input end to measure $P_\lambda(0)$ can be used. Thus, steady state condition needs to be realized in order to assign a unique value for the attenuation coefficient. A steady state launching condition can be achieved by following either the limited phase space launch technique or via insertion of a mode filter.

15.4.1 The limited phase space (LPS) launch technique

In this method, light is launched into the fiber with a restricted spatial extent and angular distribution chosen to match the steady state mode distribution as closely as possible. In this beam-optics approach as shown in Fig. 15.2, if the launch spot size (A_L, the diameter of the focused spot at the input end of the test fiber) and the launch NA satisfy the following conditions:

$$\frac{A_L}{d} = 0.707 \pm 0.5$$

$$\frac{Launch\ NA}{Fiber\ NA} = 0.707 \pm 0.05$$

where d is the core diameter of the fiber, then EMD/SMD is very nearly achieved at the launching end itself. Therefore, one needs to sacrifice only a short length (≅2–3 meters) of the fiber for measuring $P_\lambda(0)$. In this approach, lens L_2 focuses the beam on the aperture A_1 (the source aperture), which is then imaged on the input end of the test fiber with the help of aperture

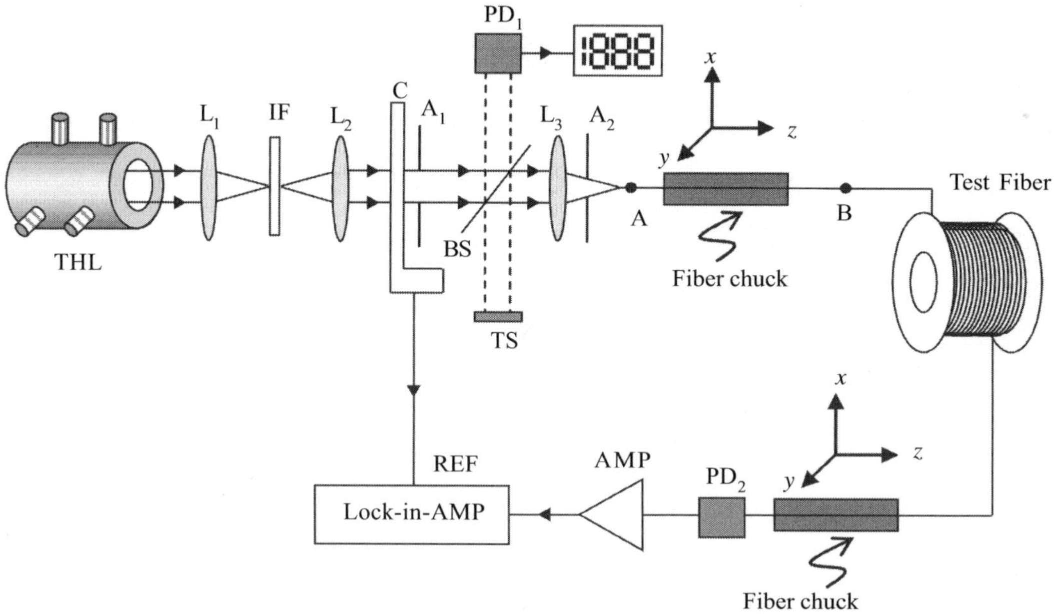

Fig. 15.2: Schematic of an experimental set-up for determination of the spectral attenuation response of a multimode fiber.

A_2 and lens L_3. Let the distance between A_1 and L_3 be l_1 and the distance between L_3 and the input face of the test fiber be l_2. Then, the launch spot size A_L will be

$$A_L = a_1 \frac{l_2}{l_1} \tag{15.2}$$

where a_1 is the diameter of the opening in A_1. The values of a_1, l_1 and l_2 can be adjusted to obtain a desired value of A_L. The opening diameter of the aperture A_2 and l_2 will obviously determine the launch NA. The opening of A_2 can be adjusted to achieve a desired value of the NA. This technique provides essentially an independent control on launch spot diameter and launch NA.

15.4.2 Mode filter launch technique

This method is based on the assumption that EMD is established in a short length of the fiber with the help of the mode filter. Test fiber is overfilled with light and a typical mode filter is realized by winding a short length of the fiber (about 5 turns) on a mandrel of diameter 1–1.5 cm (see Fig. 15.3). Such an arrangement introduces intense mode coupling and rapid attenuation of power from the higher order modes, leading to EMD. The fiber standard committees like Electronics Industries Association (EIA) recommends that "in order to qualify as an adequate mode filter, the following condition be satisfied: The far-field angle θ_s (at 5% intensity level) on a 2 m long test fiber measured with the mode filter should be within 3 ± 3% of the corresponding far-field angle (θ_L) measured on a long length of the fiber i.e. $\frac{\theta_s - \theta_L}{\theta_L} \leq 3 \pm 3\%$" [4].

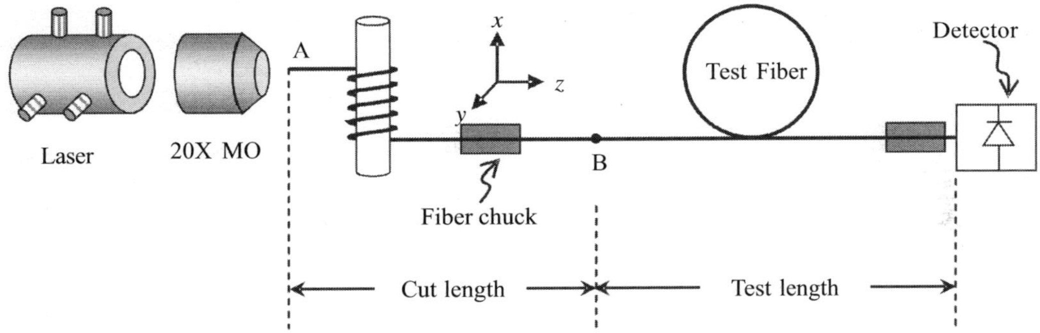

Fig. 15.3: Loss measurement using mode filter launch technique.

15.5 Procedure

Figure 15.2 shows the experimental set-up for measuring the spectral attenuation in a multimode fiber. Most part of the arrangement remains the same as in Fig. 14.2. In addition, variable apertures A_1, A_2 are used whose opening diameters can be adjusted. These together with L_3 determine the spot size and the NA of the incident beam at the input end of the test fiber. To carry out the experiment, procedure would remain the same as in the case of Experiment B of Chapter 14 and should be followed step by step.

15.6 Observations

S. No.	Length of the fiber (z) [km]	Wavelength (λ) nm	$P_\lambda(z)$ [mW]	$P_\lambda(0)$ [mW]	$\alpha_t(l)$
1. 2.					

15.7 Results

The attenuation coefficient for the given multimode fiber =

_________ dB/km @ λ_1 = _______ nm

_________ dB/km @ λ_2 = _______ nm

_________ dB/km @ λ_n = _______ nm

15.8 Comments

The main advantage of the LPS technique is the independent control over the launch conditions. However, its disadvantages are the need for bulky equipments with a precise alignment. Further, the input end-face of the test fiber has to be of high quality. The main advantage of the mode filter launch technique lies in its simplicity where bulky equipments as well as precise alignment criterion are avoided.

References

1. B.P. Pal (Ed.), *Fundamentals of Fiber Optics in Telecommunication and Sensor Systems*, New Age Publishers, New Delhi, 2002.
2. D. Gloge, "Optical power flow in multimode fibers", *Bell Sys. Tech. J.*, vol. 51, p. 1767 (1972).
3. D. Marcuse, *Principles of Optical Fiber Measurements*, Academic Press, New York (1981).
4. K. Thyagarajan, B.P. Pal and Arun Kumar, "Characterization of Optical Fibers for Telecommunication and Sensors—Part I: Multimode Fibers", *Int. J. Optoelectronics*, vol. 3, p. 45 (1988).
5. R.L. Gallawa, G.E. Chamberlain, G.W. Day, D.L. Franzen and M. Young, "Measurement of multimode optical fiber attenuation", *NBS Special Publication*, Washington, vol. 2, p. 637 (1983).
6. A.H. Cherin and W.B. Gardiner, "Standardization of optical fiber transmission measurement", Laser Focus, vol. 16, p. 60 (1980).

16

Cutoff Wavelength of a Single-Mode Fiber

Ramanand Tewari and Bishnu P. Pal

16.1 Introduction

We have seen in Chapter 4 that for a given refractive index profile, the number of guided modes depend on the core radius (a), the relative refractive index difference (Δ), and the operating wavelength (λ). These parameters can be combined into a single composite parameter called the normalized frequency V. At the operating wavelength, depending on its V-number, a fiber can be single-moded or multimoded. If the operating V-number of an optical fiber falls below a critical value, V_c, known as the normalized cutoff frequency for the LP_{11} mode, the fiber would support only the fundamental (LP_{01}) mode and it is referred to as single-mode fiber [1–3]. The precise value of V_c however depends on the core index profile shape. For a step-index fiber $V_c \cong 2.405$ while for a parabolically graded index core fiber $V_c \cong 3.5$. In practice for telecommunication applications, the refractive index profile of all single-mode fibers is nominally step-index. Thus, for single-mode operation of a step-index fiber, its V-number at the operating wavelength must lie within the range

$$0 < V < 2.405 \tag{16.1}$$

This condition can be used to obtain fiber design guidelines e.g. choice of "a" and Δ for attaining single-mode operation at a given λ. The condition (16.1) can be alternatively expressed in terms of cutoff wavelength as

$$\lambda_c = \frac{2\pi\ a\ n_1\sqrt{2\Delta}}{2.405} \tag{16.2}$$

Thus, for a given step-index fiber for which a, n_1 and Δ would be known, its λ_c can be theoretically calculated using Eq. (16.2). For wavelengths $\lambda > \lambda_c$ (of a given fiber), only the LP_{01} mode is expected to be supported by the fiber because the next higher order mode namely, LP_{11} mode and all subsequent higher order modes will be cutoff, i.e., absent in this fiber. In this context λ_c, which is essentially the cutoff wavelength of LP_{11} mode in a given fiber is a very important characteristic of the fiber since it specifies the region for single-mode operation of the fiber. Operating a fiber at a wavelength above its cutoff wavelength will ensure the propagation of only the LP_{01} mode in it. Theoretically the cutoff wavelength can be calculated if the refractive index profile of the fiber is known by solving the wave equation for the corresponding index profile; the so calculated λ_c is called the theoretical cutoff wavelength of the fiber. However, in practice, owing to various external unavoidable perturbations like macrobending, microbending, twisting etc., the fiber becomes effectively single-moded, somewhat below the theoretical λ_c [1, 3]. This happens because close to the cutoff point, the modal power in the LP_{11} mode extends deep into the cladding and therefore even a small perturbation can lead to a large loss. Thus, a fiber becomes effectively single-moded at $\lambda = \lambda_{ce}$ (< theoretical λ_c); λ_{ce} is called the effective cutoff wavelength. In fact the functional cutoff wavelength, which is measurable in practice, is always an effective cutoff wavelength. A number of methods have been proposed in the literature for measuring and defining the effective cutoff wavelength of a single-mode fiber. Because different measurement techniques at times could yield significantly different results, the ITU* has defined the effective cutoff wavelength as following:

"The cutoff wavelength (λ_c) is the wavelength such that for all $\lambda > \lambda_c$, the ratio between the total power (including launched higher order modes) and the fundamental mode power decreases to less than 0.1 dB in a quasi-straight 2 meter fiber length with one single loop of 14 cm radius".

The rationale behind this recipe can be found in Ref. 3, page 432.

For measuring the effective cutoff wavelength defined in this manner, the ITU recommends the bending technique based on spectral variation of the transmitted power through a bent fiber as reference test method. The bending technique can be implemented under two different schemes with either a single-mode fiber or a multimode fiber as the reference.

16.2 Aim

To measure the effective cutoff wavelength of a single-mode fiber.

16.3 Apparatus

Tungsten halogen lamp, monochromator, lock-in-amplifier, chopper (C), photomultiplier tube/Si- and Ge-detectors, microscope objectives, XYZ-stacks, cladding mode stripper, single-mode and multimode fibers etc.

---

* International Telecommunication Union based in Geneva, Switzerland.

16.4 Measurement principle

16.4A To measure the effective cutoff wavelength using the bend technique with single-mode fiber as reference

This technique essentially exploits the fact that a large fraction of a guided mode power resides in the cladding near the mode's cutoff wavelength. Thus, if a fiber is operated at wavelengths greater than λ_{ce}, only the LP_{01} mode in it would be guided and it will be well confined inside the core. If the operating wavelength is now decreased to a value such that the resulting *V*-number of the fiber (assuming it to be nominally step-index) becomes more than 2.405, the LP_{11} mode would be excited in the fiber. But, since it is close to its cutoff, most of its power will extend into the cladding. Hence, bending the fiber will result in high transmission loss at this wavelength. As the wavelength is decreased further, LP_{11} mode becomes increasingly more confined within the core and the transmission loss progressively decreases. Thus, there would be a peak in the loss spectrum of a test fiber near its cutoff wavelength. The cutoff wavelength will be close to that wavelength when the transmission loss starts to increase as λ is decreased beyond the pure single-mode regime.

16.4A.1 Procedure

Figure 16.1 shows the experimental arrangements used to measure the effective cutoff wavelength λ_{ce} by the ITU recommended bend technique. Collimated light from a tungsten halogen lamp (THL) is passed through a chopper (C) and focused by means of a microscope objective (L_2) on to the input end of the fiber. The output end of the fiber is held at the entrance aperture of a monochromator whose output is fed to a Si-photodetector (or a photomultiplier tube) or a cooled Ge-detector and then to a low-noise amplifier. The signal from the detector is fed to the lock-in-amplifier for phase sensitive detection. The chopper provides the reference signal for the lock-in-amplifier.

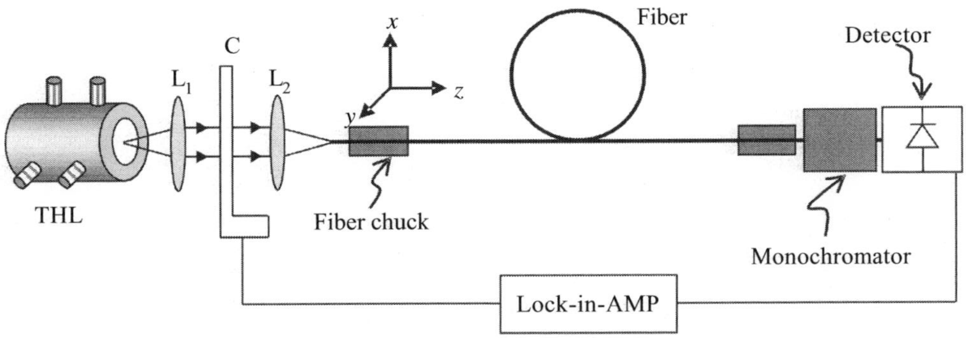

Fig. 16.1: Schematic of an experiment set-up for measurement of the effective cutoff wavelength of a single-mode fiber by the bend method.

To measure loss due to bending, first a 2 meter length of the test fiber is inserted into the apparatus and bent around a mandrel to form one loosely constrained single loop with a constant radius of 140 mm. Care should be taken to avoid any bends of radius smaller than 140 mm on the fiber lay – since the power levels involved are quite low and the cladding modes must be effectively stripped off from the fiber by inserting the cladding mode strippers immediately after and before the input and output ends of the fiber, respectively. Power transmitted through the fiber sample is then recorded by the detector as a function of wavelength λ and stored as $P_1(\lambda)$.

In the next step without changing the input launch conditions, one additional loop of sufficiently small radius is introduced into the test fiber to filter out LP_{11} mode. A typical value for the radius of this relatively tight loop is 30 mm as recommended by the ITU. Spectral transmission through the fiber sample is then measured again and stored as $P_2(\lambda)$. A measure of the bend-induced loss is then calculated as

$$R(\lambda) = 10\log\left[\frac{P_1(\lambda)}{P_2(\lambda)}\right] \tag{16.3}$$

16.4A.2 Observations

Mandrel radius =

S. No.	Wavelength (λ) nm	$P_1(\lambda)$ [mW]	$P_2(\lambda)$ [mW]	$R(\lambda)$
1.				
2.				
.				
.				

16.4A.3 Results

Figure 16.2 shows typical plot of $R(\lambda)$ for a single-mode fiber. The loop radius introduced for measurement of $P_2(\lambda)$ was 30 mm in this case. It can be seen from the figure that in a certain wavelength region the bend-induced loss indeed exhibits a peak as was expected from the explanation above. As per recommendations of ITU, a precise measure of λ_{ce} is inferred from these measurements to be *the wavelength at which the long wavelength edge of the bend-induced attenuation rises by 0.1 dB from the long wavelength base line*. This recipe, as is evident from Fig. 16.2, yield the effective cutoff wavelength λ_{ce} of the LP_{11} mode in this sample fiber to be 576 ± 4 nm. It can be further seen from this figure that there exists a second peak as λ is continued to decrease beyond λ_{ce}. This second loss peak corresponds to the effective cutoff wavelength of the next higher mode namely, LP_{02} mode. Interestingly for too small loop radius, the bend induces loss even to the LP_{01} mode at long wavelengths (as can be seen in Fig. 16.3).

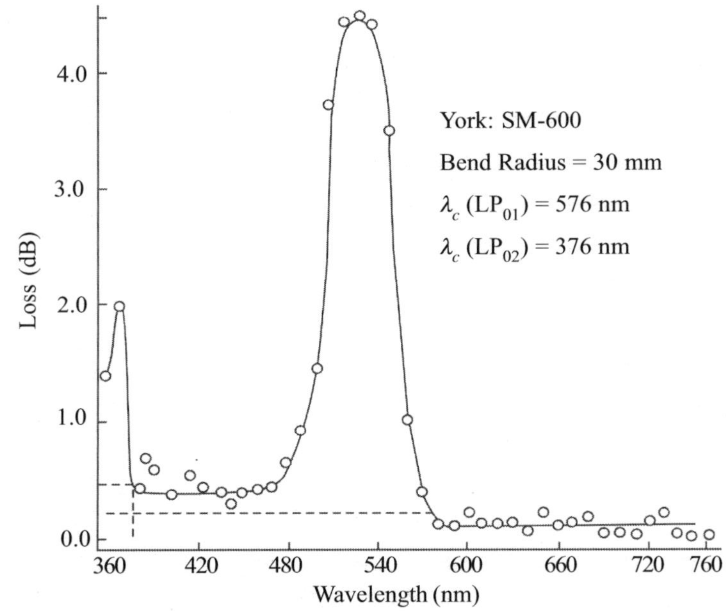

Fig. 16.2: Variation of $R(\lambda)$ given by Eq. (16.3) with λ for bend radius of 30 mm for a single-mode fiber from York Technology UK. From the figure, we obtain λ_{ce} $(LP_{11}) = 567$ nm and λ_{ce} $(LP_{02}) = 367$ nm.

Here for a very tight loop radius of 11 mm, at wavelengths where effectively only the LP_{01} mode could propagate, even the LP_{01} mode is seen to be lossy. This can be understood from the fact that at these wavelengths, the *V*-number of the fiber falls much below 2.405 and hence a large fraction of LP_{01} mode's power resides in the cladding, thereby exhibiting loss sensitivity to tight bends.

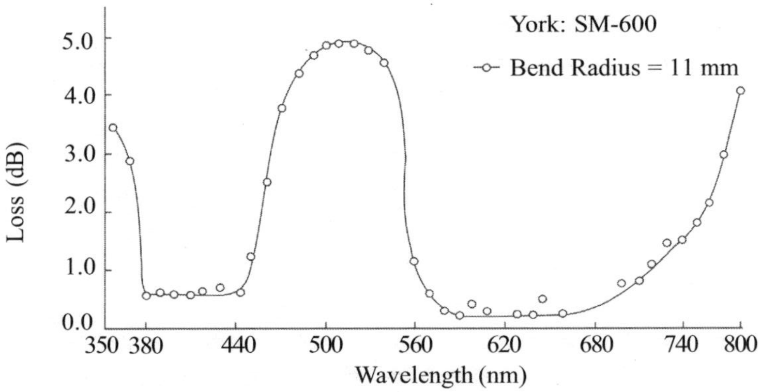

Fig. 16.3: Variation of $R(\lambda)$ with λ for bend radius of 11 mm for the same fiber as given in Fig. 16.2. The $R(\lambda)$ again starts to rise due to the bend induced loss of the LP_{01} mode.

16.4B To measure the effective cutoff wavelength using the technique with multimode fiber as reference

This method relies on the fact that optical power transmission through a single-mode fiber from a spectrally flat source would exhibit an abrupt increase at the cutoff wavelength of the LP_{11} mode due to the excitation of additional guided power through the second order mode. Since the practical sources are not spectrally flat over a broad range, and moreover the gratings, filters, detectors etc., could cause large spectral variations in system response, one normalizes the power coupled into the single-mode fiber by the power coupled into a multimode fiber, which eliminates possible large spectral variations in system response. In this method, since there is a "step" in the transmitted power versus wavelength curve, the method is also sometimes called as the "Power step" method.

16.4B.1 Procedure

The experimental set-up is the same as that of the bend method, described in Fig. 16.1. To measure the λ_{ce} by this method, first a 2 meter length of the single-mode fiber under test is inserted into the test apparatus and bent to form one loosely constrained single loop with a constant radius of 140 mm. Care should be taken to avoid any bends of radius smaller than 140 mm. As in the single-mode fiber reference method, the transmitted power $P_1(\lambda)$ is measured as a function of the wavelength λ.

In the next step, the single-mode fiber is then replaced by a similarly prepared multimode fiber and the transmitted power $P_m(\lambda)$ from the output end of the multimode fiber is measured over the same wavelength domain. In both the cases, the cladding modes must be effectively stripped off from the fiber by inserting the cladding mode strippers just after and before the input and the output ends of the fiber, respectively. The ratio between the output powers from the single-mode and the multimode fibers is calculated as

$$R(\lambda) = 10\log\left[\frac{P_m(\lambda)}{P_1(\lambda)}\right] \tag{16.4}$$

16.4B.2 Observations

Bend radius =

S. No.	Wavelength (λ) nm	$P_1(\lambda)$ [mW]	$P_m(\lambda)$ [mW]	$R(\lambda)$
1.				
2.				
.				
.				

16.4B.3 Results

Figure 16.4 shows a plot of $R(\lambda)$ for the same sample fiber from York Technology, UK. As can be seen from the figure, $R(\lambda)$ exhibits a step increase as the cutoff is approached from the long wavelength side of the curve. This is due to the additionally launched LP_{11} mode power into the fiber. As per recommendations of ITU, the effective cutoff wavelength λ_{ce} is determined to be the wavelength at which a line parallel to the long wavelength branch of $R(\lambda)$ with 0.1 dB loss difference (dashed line in Fig. 16.4) intersects the $R(\lambda)$ curve. As can be seen from Fig. 16.4, effective cutoff wavelength λ_{ce} for LP_{11} mode was found to be 578 ± 4 nm.

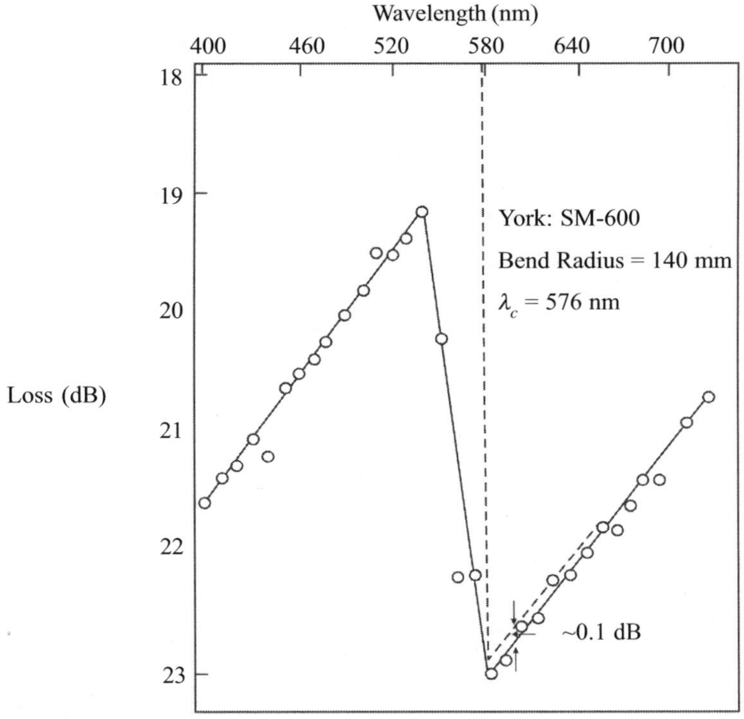

Fig. 16.4: Variation of $R(\lambda)$ given by Eq. (16.4) with λ for a single-mode fiber as given in Fig. 16.2. From the figure, we obtain $\lambda_{ce}(LP_{11}) = 578$ nm.

16.5 Comments

The effective cutoff wavelength determined by the above methods equates "cutoff" with ≅20 dB attenuation of the LP_{11} mode [3]. It is very important in telecommunication systems using single-mode fibers that the LP_{11} mode is highly attenuated before reaching splices, connectors, etc., wherein mode coupling may occur and the resultant potential modal noise may degrade the system performance. In practical applications, a separation of several kilometers is sufficient to achieve attenuation, but it can be very much less in cases of "pigtails" or in certain spliced sections. Further, in sensor applications the length of the fiber may not be more than a few meters. In such cases, a careful estimation of the cutoff wavelength, consistent with the conditions of applications, must be made. The principal advantage of the above methods is that they could be performed on the same set-up used for fiber attenuation measurements.

References

1. E.G. Neumann, *Single-Mode Fibers: Fundamentals*, vol. 57, Springer-Verlag (Berlin) (1988).
2. B.P. Pal (Ed), *Fundamentals of Fiber Optics in Telecommunication and Sensor Systems*, New Age Publishers, New Delhi (2007).
3. A. Ghatak and K. Thyagarajan, *Introduction to Fiber Optics*, Cambridge University Press, Cambridge (1998).

17

Beat Length Measurement of Birefringent Fibers

K. Thyagarajan and Ajoy Ghatak

17.1 Introduction

A perfectly circular core single-mode fiber could potentially support two modes, which have the same propagation constant but they are polarized along orthogonal directions. In practice, even telecommunication grade single-mode fiber cores may not necessarily be perfectly circular in cross-section. Furthermore, due to the possible presence of stress, twists, bends, etc. on a fiber lay such fibers are, in practice, slightly birefringent, i.e., at certain portions along its length there could exist two orthogonally polarized modes with slightly unequal propagation constants. This small difference in propagation constant leads to coupling of power with propagation between the two orthogonal polarization states even under the smallest of perturbations. Thus, for nominally circular core fibers it is found that a linearly polarized incident light beam quickly becomes elliptically polarized after propagating over a short distance in the fiber. In addition, this output state of polarization is sensitive to changes in temperature, wavelength or any other external perturbation. For applications of single-mode fibers in coherent communications, interferometric sensors, coupling to polarization-sensitive integrated optic devices, etc. this randomly changing state of polarization could cause problems. In order to cater to such applications, new fiber designs have been developed, which are fabricated to achieve large linear birefringence at the production stage itself, so that due to inherent birefringence that would exist in such fibers small perturbations cannot cause coupling of power from one polarization to another. In such high birefringent fibers (referred to as polarization maintaining fibers), light launched into any one of the birefringence axes will remain in that state under normal external perturbations.

There are mainly two types of high birefringence fibers namely elliptic core (EC) fibers and stress induced birefringent (SIB) fibers (see Fig. 17.1). In the former, although the fiber is made of isotropic material, the elliptic shape of the core leads to birefringence. This type of birefringence is also referred to as geometrical birefringence. In SIB fibers, two stress producing members are introduced on either side of the core which induces a birefringence in the core through the photoelastic effect. This birefringence of SIB fibers is usually much higher than that of EC fibers. It is comprehensible that a vertically polarized light would effectively experience a fiber cross-section either in dimension (in an EC fiber) or in refractive index distribution (in case of an SIB fiber) different from that of a horizontally polarized light and hence the effective mode index experienced by the two orthogonally polarized propagating lights in such fibers would be different. In this chapter we will describe an experiment to measure the birefringence of a highly birefringent fiber through observation of Rayleigh scattering [1, 2].

17.2 Aim

Measurement of beat length of birefringent single-mode fibers using Rayleigh scattering technique.

17.3 Apparatus

633 nm He-Ne laser of at least 5 mW output power, half wave plate, quarter wave plate, high birefringent fiber, traveling microscope, index matching oil, 20X MO, XYZ-translation stages.

17.4 Evolution of state of polarization in a high birefringent fiber

The elliptic core fiber (see Fig. 17.1) is perhaps the simplest example of a birefringent fiber; the "modes" of the fiber are (approximately) *x*- and *y*-polarized; i.e., if an *x*-polarized beam is incident, it will propagate without any change in the state of polarization with a certain phase velocity $\dfrac{\omega}{\beta_x}$. Similarly, a *y*-polarized beam will propagate as a *y*-polarized beam with a phase velocity $\dfrac{\omega}{\beta_y}$. Now, let us assume that a left circularly polarized beam be incident on the input face of the fiber at $z = 0$; in order to study propagation of the beam through the fiber, we must resolve this incident beam into *x*- and *y*-polarized beams propagating with slightly different velocities. Thus

$$\mathbf{E}(x, y, z, t) = A\psi(x, y)\,[\,\hat{\mathbf{x}} \cos(\omega t - \beta_x z) - \hat{\mathbf{y}} \cos(\omega t - \beta_y z)] \tag{17.1}$$

where $\psi(x, y)$ is the normalized transverse field distribution of the fundamental mode and A is a constant determined by the power carried by the mode. The circularly polarized input excites equally the x- and y-polarized modes. For a circular core fiber, $\beta_x = \beta_y$ and the beam remains circularly polarized for all values of z. Now, at $z = 0$

$$E_x = A\psi(x, y) \cos \omega t, \quad E_y = -A\psi(x, y) \sin \omega t \tag{17.2}$$

which represents a left circularly polarized wave (see Fig. 17.2). For

$$z = z_1 = \frac{\pi}{2(\beta_y - \beta_x)} \tag{17.3}$$

i.e., for $\quad \beta_y z_1 = \beta_x z_1 + \dfrac{\pi}{2}$

$$E_x = A\psi(x, y) \cos(\omega t - \phi_1)$$

$$E_y = -A\psi(x, y) \sin\left(\omega t - \phi_1 - \frac{\pi}{2}\right) = +A\psi(x, y) \cos(\omega t - \phi_1) \tag{17.4}$$

where $\phi_1 = \beta_x z_1$. Equation (17.4) represents a linearly polarized wave polarized at 45°

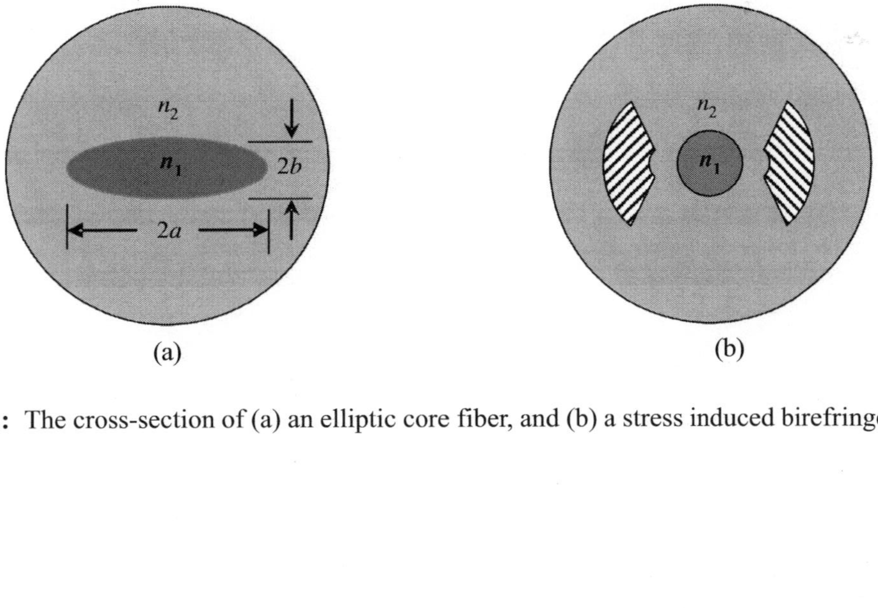

Fig. 17.1: The cross-section of (a) an elliptic core fiber, and (b) a stress induced birefringent fiber.

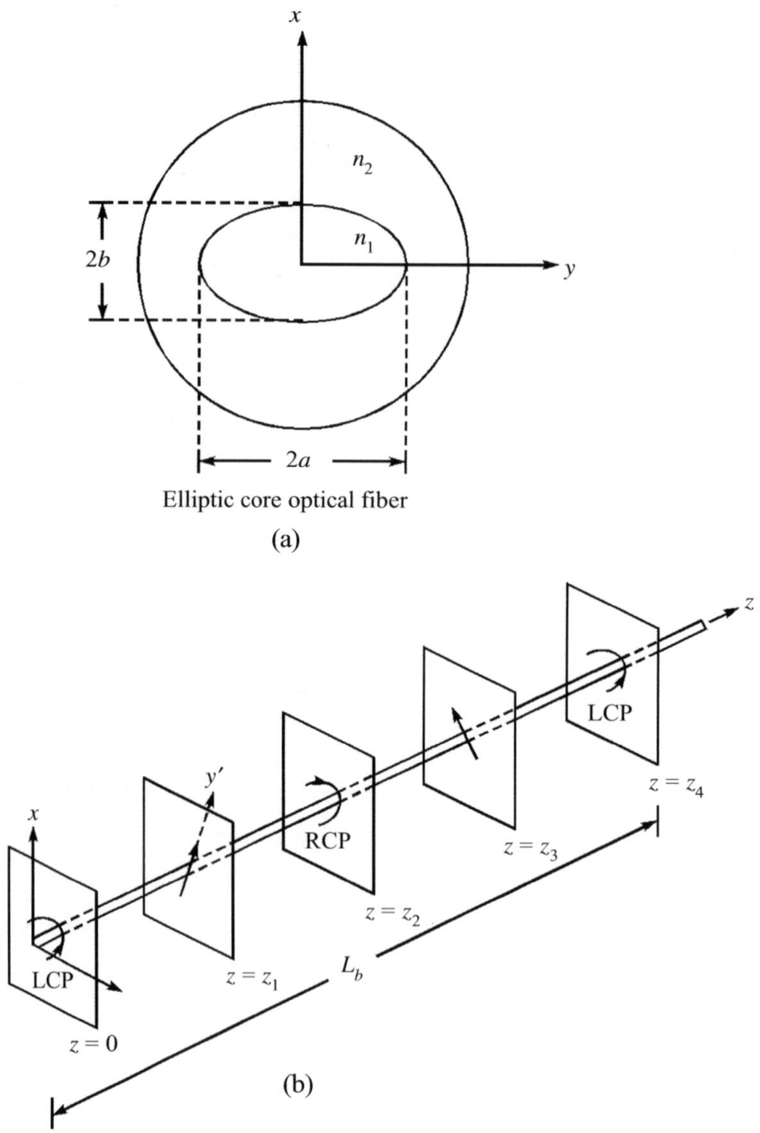

Fig. 17.2: (a) The modes of an elliptic core are (approximately) x-polarized and y-polarized. (b) Evolution of the SOP of a left-circularly polarized beam incident on an elliptic core fiber. If we view along the y'-axis then dark spots will be observed at $z = z_1, 5z_1, 9z_1, \ldots$; figure adapted from Ref. 3.

to the x (or y) axis (see Fig. 17.2); we assume the direction of the **E**-vector to be along the y' axis. Similarly, at

$$z = z_2 = \frac{\pi}{(\beta_y - \beta_x)} = 2z_1$$

$$E_x = A\psi(x, y)\cos(\omega t - \phi_2)$$

$$E_y = -A\psi(x, y)\sin(\omega t - \phi_2 - \pi) = +A\psi(x, y)\sin(\omega t - \phi_2) \tag{17.5}$$

where $\phi_2 = \beta_x z_2$

and the wave will be right circularly polarized (see Fig. 17.2). At

$$z = z_3 = \frac{3\pi}{2(\beta_y - \beta_x)} = 3z_1$$

we will have $E_x = A\psi(x, y)\cos(\omega t - \phi_3)$

$$E_y = -A\psi(x, y)\sin\left(\omega t - \phi_3 - \frac{3\pi}{2}\right) = -A\psi(x, y)\cos(\omega t - \phi_3) \tag{17.6}$$

where $\phi_3 = \beta_x z_3$

Thus, the wave would again be linearly polarized but now the direction of the oscillating electric field will be at right angles to the field at $z = z_1$. In a similar manner, we can easily continue to determine the SOP of the propagating beam. Thus, at $z = 5z_1, 9z_1, 13z_1, \ldots$ the SOP will be the same as at $z = z_1$ and at $z = 7z_1, 11z_1, 15z_1, \ldots$ the SOP will be the same as at $z = 3z_1$. Similarly, at $z = 4z_1, 8z_1, 12z_1, \ldots$ the beam will be LCP and at $z = 2z_1, 6z_1, 10z_1, \ldots$ the beam will be RCP.

From the above discussion it follows that the SOP at any value of z repeats after every propagation distance of $4z_1$; this distance is called the *beat length* L_b of the fiber. Hence

$$L_b = 4z_1 = \frac{2\pi}{|\beta_y - \beta_x|} \tag{17.7}$$

The birefringence of a fiber is measured in terms of the difference in effective indices of the two orthogonally polarized modes of the fiber. Thus, if x and y are the eigen axes of the fiber and β_x, and β_y the corresponding propagation constants then we define the effective indices as

$$n_x = \frac{\beta_x}{k_0}, \; n_y = \frac{\beta_y}{k_0} \tag{17.8}$$

where $k_0 = \dfrac{2\pi}{\lambda_0}$ and λ_0 is the free space wavelength. The birefringence of the fiber is defined as

$$B = |n_x - n_y| \tag{17.9}$$

Thus,

$$L_b = \frac{\lambda_0}{B} \tag{17.10}$$

As a numerical example, we consider an elliptic core fiber for which

$$2a = 2.14\ \mu m, \qquad 2b = 0.85\ \mu m$$

$$n_1 = 1.535, \qquad n_2 = 1.47$$

(see Fig. 17.1). For such a fiber operating at $\lambda_0 = 0.6328\ \mu m$ ($k_0 \approx 9.929 \times 10^4$ cm^{-1})

$$n_x = 1.506845 \text{ and } n_y = 1.507716$$

The beat length is given by $L_b = \dfrac{\lambda_o}{|n_x - n_y|} \cong 0.727$ mm

The beat length is a good measure of the birefringence of the fiber. Larger the birefringence, smaller will be the beat length. Thus, high birefringence fibers are characterized by a small beat length. A typical high birefringence fiber would have a beat length of a few millimeters. A beat length of 3 mm at a wavelength of 0.6 μm corresponds to a birefringence of 2×10^{-4}. Compare this with the bulk birefringence of quartz, which has a value of 1×10^{-2} and that of lithium niobate, which has a value of 9×10^{-2}. Thus, the induced birefringence in fibers is very small compared to naturally occurring birefringences in crystals. Even such small birefringence can very strongly induce coupling of light from one polarization to another.

17.4.1 Rayleigh scattering

Rayleigh scattering is a fundamental mechanism which scatters light out of the fiber [3]. This scattering is produced by the frozen density and dopant concentration fluctuations in the core of the fiber over dimensions smaller than the wavelength. This scattering decreases as the inverse fourth power of the wavelength. This is one of the reasons why today's optical fiber communication systems operate at the longer wavelength of 1.55 μm rather than at the first generation wavelength of 0.8 μm wavelength. One can easily check the wavelength dependence of Rayleigh scattering by coupling white light from, say, a tungsten halogen lamp into a long (200 m or more) multimode fiber and observe the colour of the emerging light with naked eyes. The emerging light would be reddish in colour implying greater scattering at lower wavelengths. (Recollect the reasons for a blue sky and a reddish rising or setting sun).

It is well known that for a linearly polarized light undergoing scattering, there is no scattered wave along the axis of the dipole, i.e., along the direction of polarization and the scattering is maximum in a direction perpendicular to the dipole axis. It is this effect which is used in the measurement of the beat length of fibers.

Returning to the example discussed earlier, let the fiber be rotated in such a way that the y' axis is along the vertical line (the x' and the z axes are assumed to lie in the horizontal plane). Thus, if we place our eyes vertically above the fiber and view vertically down, then the regions $z = z_1$, $5z_1$, $9z_1$, ... will appear dark (see Figs.17.3 and 17.4). This is because of the fact that in these regions, the electric field is oscillating in the y' direction (which is the vertical direction) and we know that if the dipole oscillates along the y' direction, there is no radiation emitted in that particular direction. Thus, by measuring the distance L_b between two consecutive dark regions one can calculate $\beta_y - \beta_x$. Furthermore, by moving the eyes to the horizontal plane, i.e., viewing along the x' axis, we will see the regions $z = z_1, 5z_1, 9z_1, \ldots$ appear bright, and the regions $z = 3z_1$, $7z_1$, $11z_1$, ... appear dark. Thus, the experiment not only allows one to understand the changing SOP of a beam propagating through a birefringent fiber, but it also helps us in understanding the radiation pattern of an oscillating dipole.

17.5 Procedure

The experimental set-up for the measurement of beat length of a birefringent single-mode fiber is shown in Fig. 17.5. The following procedures have to be carried out step by step:

1. Fiber ends are prepared so that it has well-cleaved flat ends.
2. In order to excite both the eigen modes of the fiber with equal amplitude, light from a linearly polarized (LP)[1] He-Ne laser should be passed through a quarter wave plate (QWP) before coupling to the fiber to convert the state of polarization of the incident light into a circular state of polarization. One could equally well have used the linearly polarized laser directly; however one would then have to align the orientation of the polarization of the input beam at 45° to the eigen axes of the fiber so as to excite both the eigen modes with equal amplitude. On the other hand, use of a circular polarization does not require any such intricate alignment.
3. To couple the light into the fiber, laser is then focused on the entrance face of a well cut fiber using a 20X microscope objective.

[1] One could as well use an unpolarized laser, but one would then have to pass it through a polarizer (and if necessary a quarter wave plate) before coupling into the fiber. This would of course result in a loss of at least 50% of light. Why?

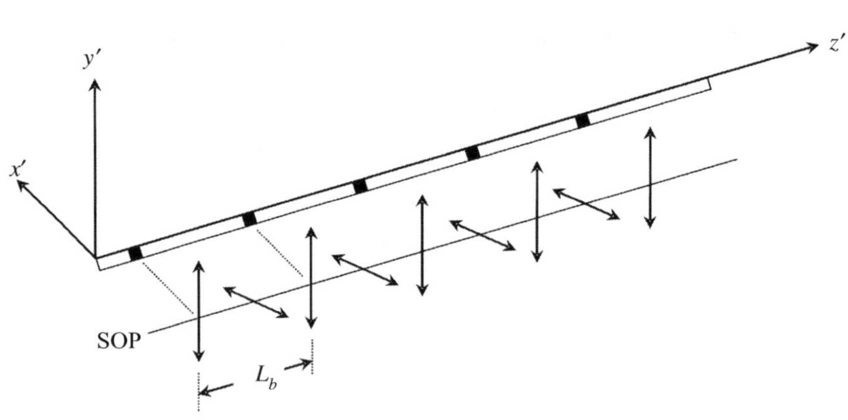

Fig. 17.3: If linearly polarized light, polarized at 45° to the eigen axes is incident on the fiber, then due to the evolution of the SOP, one can observe a beat pattern by observing the fiber at 45° to the eigen axes.

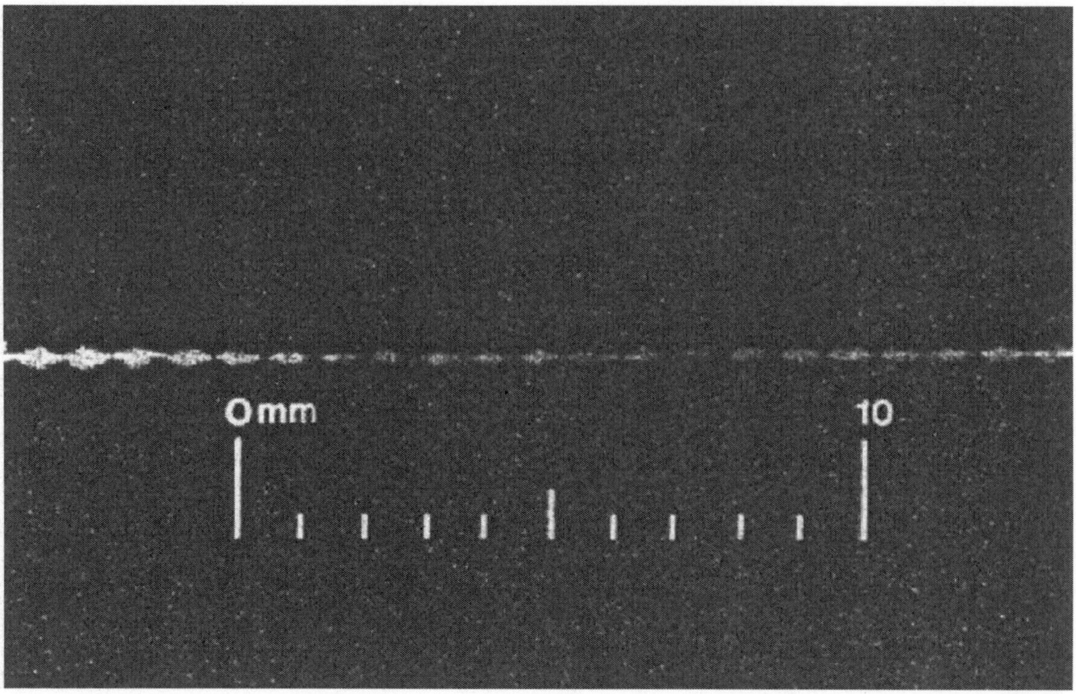

Fig. 17.4: A circularly polarized beam is incident normally on an elliptic core fiber—the photograph shows the intensity variation as seen from the top; adapted from a photograph by Andrew Corporation, Orland Park, IL, USA; the authors came across this photograph in Ref. 6.

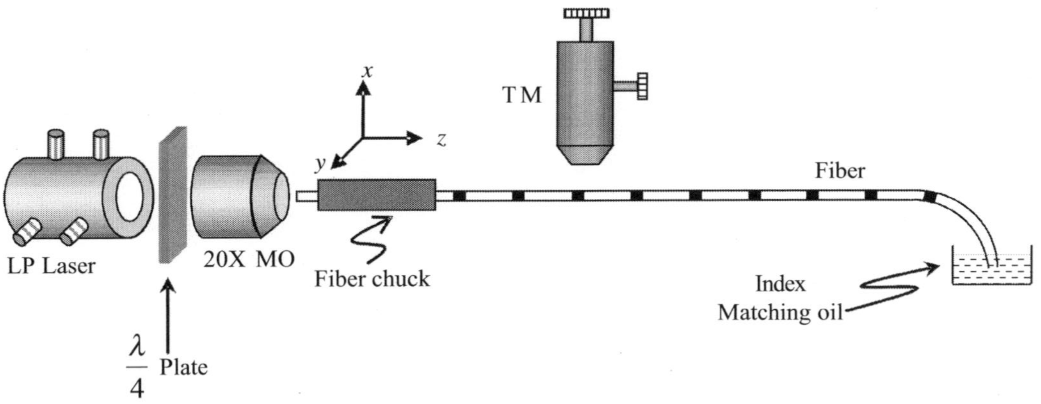

Fig. 17.5: Experimental set-up to measure the beat length of a high birefringence fiber using Rayleigh scattering technique.

4. Since the scattering is usually very weak, in order to be able to see the beat pattern very clearly, it is necessary to remove the cladding modes using a cladding mode stripper. Also since a fraction of the light propagating in the fiber may be re-coupled into the fiber from the output end, and which in turn, will create its own beat pattern (not necessarily in phase with the existing one), the output end should be dipped in an index matching oil to remove/reduce back reflection.
5. The following important observations will confirm that the observed pattern is due to the change of state of polarization:
   (a) After coupling in circularly polarized light, if one observes the fiber in a direction along any one of the eigen polarizations, then no beat pattern will be observed.
   (b) If one couples linearly polarized light, polarized along one of the eigen polarizations, then again no beat pattern will be observed irrespective of the direction of observation.
6. Once a good coupling has been accomplished, the fiber is placed in the field of view of a traveling microscope (TM) with its eigen axes at 45° to the viewing direction. One can achieve this by rotating the axis of the fiber so as to obtain the maximum contrast in the beat pattern. Using the traveling microscope one then measures the separation between say, 10 bright regions from which the average beat length can be calculated.

17.6 Observations

S. No.	Distance between 10 intensity maxima, L (mm)	Beat Length ($L_b = L/10$) mm
1.		
2.		
.		

17.7 Results

Average beat length (L_b) =

Birefringence of the fiber $\left(B = \dfrac{\lambda_0}{L_b}\right) =$

17.8 Comments

This experiment works well with small beat length or high birefringence fibers. Typically the beat length should be around a few millimeters for this experiment. Also the smaller the wavelength the larger will be the scattering amplitude and the better will be the beat pattern. Thus, measurement at green wavelength will result in larger scattering amplitude than at the red wavelength. At the same time one must remember that the fiber should be single-moded at the measurement wavelength so that the contrast is good. Otherwise there would be beat patterns corresponding to separate spatial modes which will be non-overlapping since the beat length of each of those modes will be different.

The Rayleigh scattering technique can also be used for the measurement of the twist induced circular birefringence in fibers. If a nominally circular core single-mode fiber is given a twist, then it can be shown that the fiber becomes circularly birefringent, i.e., the modes of the fiber becomes right and left circularly polarized [4]. In such fibers, a circularly polarized incident wave will remain in that state as it propagates in the fiber. If linearly polarized laser is coupled into the twisted single-mode fiber, then due to the circular birefringence the plane of polarization of the beam will rotate as it propagates through the fiber and the state of polarization will remain linear. If one now observes the fiber from the side, then one would again see a beat pattern due to the rotating nature of the linear polarization. By measuring the distance between the beats, one can again measure the twist induced circular birefringence. One could indeed measure the increase in induced circular birefringence as the fiber is twisted at higher and higher rates [5].

To have a deep insight and understanding, one can further carry out the following experiments:

1. Show experimentally and analytically that maximum contrast is observed if the two modes are equally excited.
2. Make an arrangement to fix the ends of the fiber so that the fiber can be twisted and held in this condition. Devise an experiment to measure the twist induced circular birefringence in the fiber.
3. Couple light into one of the eigen polarization states and show that the output state of the polarization does not change with nominal perturbation such as bending, pressing, temperature change etc. Perform the experiment with an ordinary single-mode fiber and compare with the results obtained with high birefringence fibers.

4. Couple an unpolarized laser beam into the fiber and see if any beat pattern is observed. Explain your observations.

References

1. A. Papp and H. Harms, "Polarization optics of liquid-core optical fibers", *Applied Optics*, vol. 16, p. 1315 (1977).
2. K. Thyagarajan, B.P. Pal and Arun Kumar, "Characterization of optical fibers for telecommunication and sensors: Part II, Single-mode fibers", *Int. J. Optoelectronics*, vol. 3, p. 153 (1989).
3. Ajoy Ghatak, *Optics*, 4th Edition, Tata McGraw-Hill, New Delhi (2008).
4. R. Ulrich and A. Simon, "Polarization optics of twisted single-mode fibers", *Applied Optics*, vol. 18, p. 2241 (1979).
5. P. Graindorge, K. Thyagarajan, H. Arditty and M. Papuchon, "Scattered light measurement of the circular birefringence in twisted single-mode fiber", *Optics Comm.*, vol. 41, p. 164 (1982).
6. L.B. Jeunhomme, *Single-Mode Fiber Optics*, Marcel Dekker, New York (1983).

18

Selective Excitation of Guided Modes in an Optical Fiber

K. Thyagarajan

18.1 Introduction

In Chapter 4, we had seen that the guided modes of an optical fiber form a discrete set and that each mode is characterized by a well defined transverse field distribution and a definite propagation constant. For example, Fig. 18.1 shows the transverse intensity patterns of the LP_{01}, LP_{11} and LP_{21} modes [1].

In most of the experiments on optical fibers, modes are excited by focusing a laser beam at the entrance face of the fiber. If the fiber is single-moded then of course, the incident light will excite the only possible mode namely the LP_{01} mode. On the other hand, if the fiber can support more than one mode, then using such an excitation technique it is very difficult to excite the modes individually and one has to use some other method which will discriminate among the modes. Such a technique is the prism coupling method, which we will now discuss.

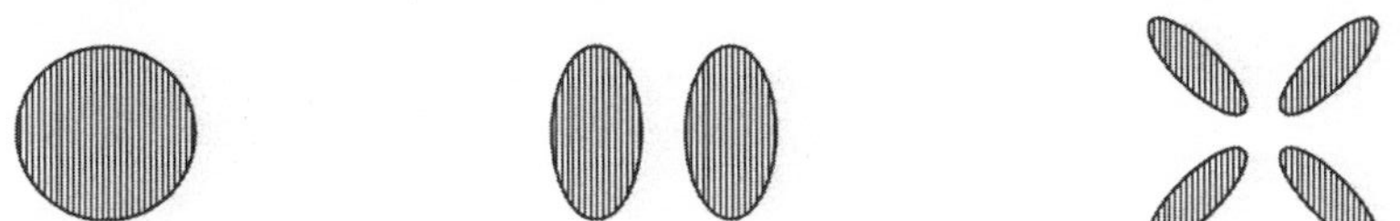

Fig. 18.1: Schematic representation of the transverse intensity patterns corresponding to the LP_{01}, LP_{11} and LP_{21} modes of a circular core step-index fiber.

In Fig. 18.2 we have shown the transverse section of a set-up consisting of a waveguide and the prism. The prism, whose refractive index is higher than that of the core of the waveguide, is placed on the top of the waveguide. Light from a laser is made to be incident on the base of the prism at an angle greater than the critical angle for total internal reflection at the interface between the prism and air. In Chapter 4, we saw that a guided mode can indeed be considered as a superposition of plane waves propagating in the core of the waveguide and undergoing total internal reflection at the two waveguide interfaces. Since a mode represents a standing wave pattern in the transverse direction, it is only at certain discrete angles of incidence of the corresponding plane waves that the total transverse phase shift in one round trip will be an integral multiple of 2π. (It is similar to the case of standing waves on a string wherein two oppositely propagating waves in the string interfere to produce the standing wave pattern. In order for this to happen, the phase shift in one round trip must be an integral multiple of 2π. It is this condition which leads to the discrete spectrum of frequencies of oscillation of the string.) For an arbitrary angle of incidence of the wave on the prism base, waves will try to enter the waveguide core by the phenomenon of frustrated total internal reflection (FTIR). The angle at which these waves are propagating in the core of the waveguide will depend on the refractive index of the prism and the core of the waveguide and also on the angle of incidence on the prism base. If the angle of propagation of the waves in the core of the waveguide does not correspond to that of a guided mode then the waves will destructively interfere in the core and no build up of the mode energy in the waveguide will take place. On the other hand, if the angle corresponds to that of a guided mode then the incoming waves will interfere constructively within the core of the waveguide and the energy in the waveguide builds up. Beyond the position of the prism, the light energy which has entered the waveguide from the prism base propagates as a mode. Now, since the propagation constant of different guided modes are different (and hence the angle of propagation of the component plane waves is different for different modes), one can excite different modes individually by letting the light wave to be incident on the base of the prism at different angles. It is this concept which will be used to excite individual guided modes in an optical fiber.

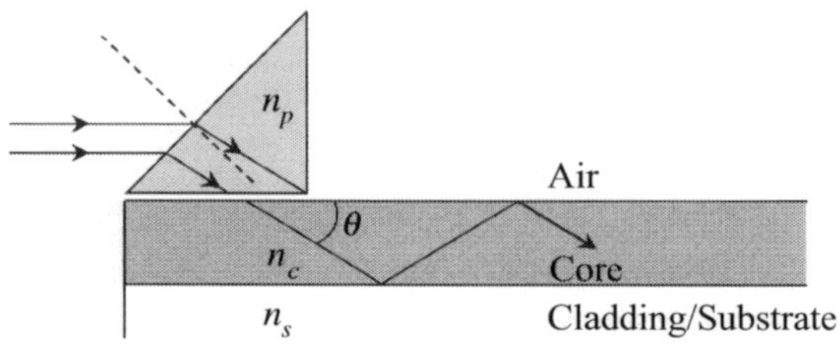

Fig. 18.2: (a) Coupling of light into a waveguide using the prism coupling arrangement through FTIR.

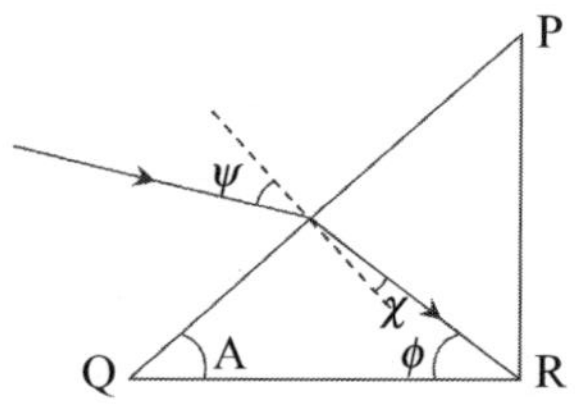

Fig. 18.2: (b) Definition of relevant angles in the experiment on coupling of light into a waveguide using the prism coupling arrangement through FTIR.

18.2 Aim

Demonstration of the excitation of guided LP modes in an optical fiber using the prism coupling technique and measurement of propagation constants of various modes.

18.3 Apparatus

Polished fiber half block, laser, microscope objective, fiber positioning set-up, prism, spectrometer table, long focal length (15–20 cm) lens.

18.4 Theory

As discussed earlier, in order to excite a guided mode in the waveguide, the angle of incidence ϕ [see Fig. 18.2(b)] must be such that the waves, which enter the waveguide core by the phenomenon of FTIR should be propagating at an angle corresponding to the mode. Applying Snell's law on the surface QR, we find that the angle of propagation θ of the waves entering the core will be such that

$$n_c \cos\theta = n_p \cos\phi \tag{18.1}$$

Now, referring to Chapter 4, we note that the plane wave components of a mode with propagation constant β travel at an angle given by

$$\beta = k_0 n_c \cos\theta_0 \tag{18.2}$$

Thus, for excitation of a guided mode we must have $\theta = \theta_0$ and we get

$$\beta = k_0 n_p \cos\phi \tag{18.3}$$

or

$$n_p \cos\phi = \frac{\beta}{k_0} = n_{\text{eff}} \tag{18.4}$$

where n_{eff} is the effective index of the mode and n_p is the refractive index of the prism. Since $\cos\phi \le 1$ and $n_{\text{eff}} \le n_c$, it can be seen from Eq. (18.4) that in order to be able to excite all guided modes, $n_p > n_c$ i.e., the prism refractive index should be greater than that of the core.

In order to relate n_{eff} to the angle ψ in Fig. 18.2, we first use Snell's law on the surface PQ

$$\sin\psi = n_{\text{p}} \sin\chi \tag{18.5}$$

Now from Fig. 18.2(b), $A + \phi + \dfrac{\pi}{2} + \chi = \pi$

or

$$\phi = \frac{\pi}{2} - \text{A} - \chi \tag{18.6}$$

where χ is the angle of refraction inside the prism and A is the angle of the prism [see Fig. 18.2(b)]. Thus

$$n_{\text{eff}} = n_p \cos\left(\frac{\pi}{2} - A - \chi\right)$$

$$= n_p \sin\left[A + \sin^{-1}\left(\frac{\sin\psi}{n_p}\right)\right] \tag{18.7}$$

Hence, by experimentally measuring the angle ψ when coupling to a particular mode takes place, we can obtain n_{eff} of the mode. Also since different guided modes have different n_{eff} values, the corresponding coupling angles ψ will be different. Thus, we can individually excite different guided modes of the fiber using this technique. This technique is extensively used in the field of integrated optics for coupling light into planar waveguides and for characterizing them [2].

We know that for a waveguide with a core index n_1 and cladding index n_2, the effective indices of all guided modes lie between n_1 and n_2. Since the maximum value of the sine function is unity, from Eq. (18.7) we can see that for coupling light into the waveguide, the refractive index of the prism must be greater than the refractive index of the core of the waveguide. In addition, since we are using the frustrated total internal reflection phenomenon for coupling, the separation between the prism base and the core of the waveguide should be small and of the order of the wavelength of light.

We have studied in Chapter 4 that in an optical fiber the core is surrounded by a cladding of lower refractive index. Further, the cladding thickness is so chosen that the field of the mode is negligible at the cladding air interface. In usual fibers, the cladding thickness is about 35 to 40 μm. Thus, if we place a prism on the cladding of the fiber for exciting the mode, then due to the large thickness of the cladding, no light will be able to penetrate the core due to FTIR and hence the guided mode cannot be excited. Hence to use this technique for optical fibers, we must first remove a major portion of the cladding from one side of the fiber as shown in Fig. 18.3 and then place the prism on the flat surface for coupling light. This is accomplished by using the fiber side polishing technique [see e.g., Ref. 3]. Briefly, the fiber is placed on a groove made on a quartz block and polished from the side until 1 to 2 micrometers from the core-cladding interface is reached. In addition to bringing the prism close to the fiber core, the quartz block also provides a base for placing the prism for coupling.

18.5 Procedure

18.5.1 Excitation of guided LP modes

The experiment can be performed in two ways – (i) light can either be coupled into the fiber with the help of the prism, in which case, the end-face of the fiber would be observed to see what kind of modes has been excited or, (ii) light can be coupled into the end-face of the fiber using appropriate microscope objective. In the latter case, the prism is used to couple light out of the fiber. Former case enables us to see different transverse mode patterns of the fiber, whereas the latter case enables us to see that the light corresponding to different modes come out of the prism along different directions – signifying the discrete nature of the guided modes. Figures 18.3 and 18.4 show the experimental set-ups for excitation of different guided modes of the fiber for the methods (i) and (ii), respectively. To carry out the experiments, the following procedures have to be executed step by step:

1. Fiber ends of polished half-block are prepared so that it has well-cleaved flat ends.
2. Place a prism having a typical refractive index of 1.5 or so on a polished fiber half-block.

   Since most of the telecommunication optical fibers are silica-based fibers with a core refractive index about 1.47, a prism of refractive index 1.5 would be suitable. Prisms having still higher refractive index can also be used in the experiment. Further, in order to see various guided LP modes, fiber capable of supporting 3 to 4 modes at the wavelength of available laser source should be chosen. Indeed, a single-mode optical fiber designed for the operating wavelength around 1.3 μm, will be able to support about 4 modes at 633 nm wavelength of commonly available He-Ne laser. Thus, a polished half block with such a fiber will be very appropriate for this experiment.

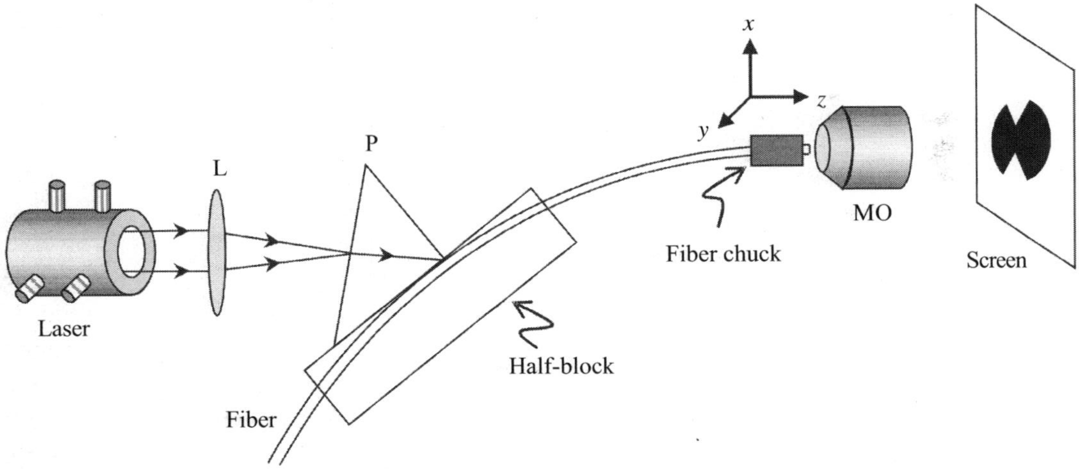

Fig. 18.3: Experimental arrangement for coupling individual LP modes of the fiber using the prism coupling set-up.

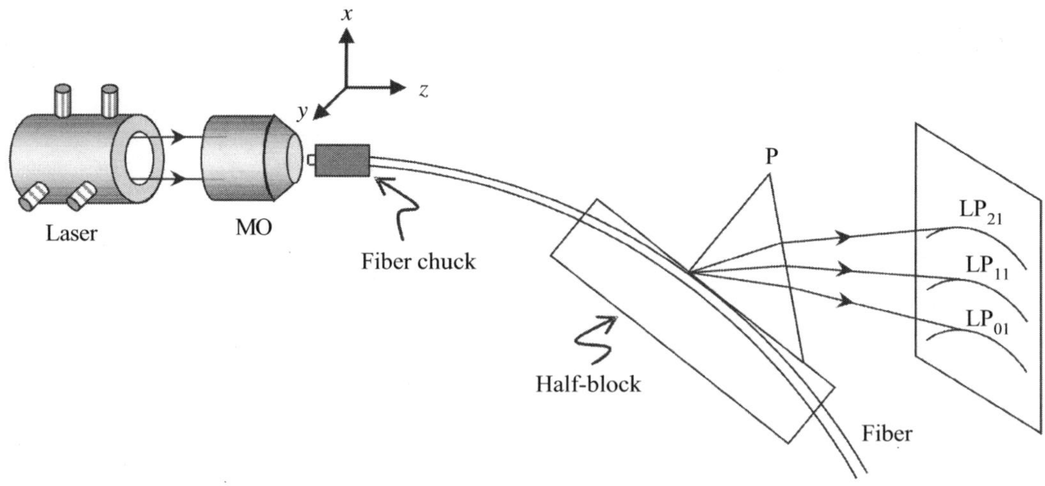

Fig. 18.4: Experimental arrangement to observe relative powers in various modes.

3. In order to see the transverse mode patterns corresponding to different modes, place the fiber half-block on a set-up which can be rotated about a horizontal axis using a goniometer arrangement. In absence of such a set-up, one can equally well use a spectrometer table and place the half-block with the prism on the prism table in a vertical position. In both the above arrangements, align the source such that the laser beam is incident on the prism base at different angles and thus excites different modes of the fiber. The set-up should also have arrangements for transverse movements so that the laser beam could be made incident on the polished region of the fiber.

4. Image the output end of the fiber on a screen, kept at a distance of about a meter from the microscope objective to get a large magnification. Allow the beam from the laser to fall on the polished region of the fiber through the prism, and rotate the prism-fiber combination to change the angle of incidence on the base of the prism. If one observes the magnified image of the fiber core as one changes the angle of incidence, one will see the appearance of different mode patterns on the screen. As one decreases the angle of incidence ψ from large values, the first pattern to appear will be that of the LP_{01} and will appear as shown in Fig. 18.1(a). The pattern will look like that of a Gaussian beam and will have no zeroes in intensity. The next to appear will be the LP_{11} whose intensity distribution will look like that shown in Fig. 18.1(b). As one further decreases, subsequent modes keep appearing one at a time as sequenced in Fig. 18.1. These observations indicate to the following:

   i. The guided modes of the optical fiber form a discrete set.

   ii. Different modes have different values for the propagation constant.

   iii. The fundamental mode has the largest propagation constant and it also has a Gaussian-like field distribution.

18.5.2 Measurement of propagation constants of various modes

The following steps describe the procedure to obtain the effective indices corresponding to different modes of the fiber:

1. Fix the polished fiber half-block with the prism on top on a spectrometer table in the vertical orientation so that the angle ψ can be changed. If a goniometer is available, it would be more convenient since the axis of rotation will be horizontal and placing the prism on the polished block will be easier.
2. Align a laser beam and a long focal length (15~20 cm) lens so that the beam is almost focused on the bottom surface of the prism (see Fig. 18.3).
3. Rotate the prism table with the polished half-block so that the beam reflected by the hypotenuse of the coupling prism goes back almost into the laser. This position will correspond to the laser beam incident normally on the prism hypotenuse. Note down the angle in the spectrometer.
4. Now rotate the prism so as to increase ψ (clockwise in Fig. 18.3) while observing the output from the fiber end and adjust it to obtain maximum coupling into each individual fiber mode. Note all the corresponding angles. From these angles and the angle measured earlier in step 3, obtain the corresponding ψ values.
5. From the measured ψ values and knowing n_p and A calculate the effective indices using Eq. (18.7).
6. If the refractive index profile of the fiber is known, you may compare the measured n_{eff} values with that estimated theoretically.

18.6 Observations

Prism refractive index =

Fiber core refractive index =

Prism angle =

Angle for normal incidence =

S. No.	Angle ψ (degree)	n_{eff}
1. 2.		

18.7 Results

Propagation constant for LP_{01} mode = ...

Propagation constant for LP_{11} mode = ...

18.8 Comments

To get a deeper insight and understanding, one can further carry out the following experiments:

1. In the arrangement shown in Fig. 18.4, couple in white light and observe the light emanating from the prism. Explain your observations.
2. The relative powers of the various beams coming out of the prism in the arrangement of Fig. 18.4 correspond approximately to the relative powers carried by various modes of the fiber. Try bending the fiber in the region between the input and the prism and notice how the relative powers of the beams exiting from the prism change. The power in the higher order modes will reduce much more than the power in the fundamental mode. Why? Try changing the input conditions and notice how the relative powers in various modes change.
3. Using the set-up of Fig. 18.3, measure the transverse extent of various modal patterns and draw conclusions regarding variation of mode size with mode number.

References

1. J.E. Midwinter, *Optical Fibers for Transmission*, John Wiley and Sons, New York (1979).
2. T. Tamir, "Beam and waveguide couplers", in *Integrated Optics*, T. Tamir (Ed.), Springer-Verlag, Berlin (1975).
3. A.C. Boucouvalas, "Fiber optic components", in *Principles of Modern Optical Systems*, I. Andonovic and D. Uttamchandani (Eds.), Artech House Inc., Norwood (1989).

19

Study of Mode-Mode Fiber Interferometer: Application to Temperature Sensing

R.K. Varshney and Arun Kumar

19.1 Introduction

Fiber optic sensors, which can be used successfully to sense parameters like temperature, pressure, strain etc., offers many advantages (e.g., high sensitivity, reliability, remote sensing, and immunity to ambient electrical noise, high voltage and electromagnetic interference etc.) over existing techniques. Interferometric sensors, which are based on the phase modulation by the sensing parameter, are among the most sensitive fiber optic measurement devices. These sensors generally involve two single-mode fibers, one of which is used as reference arm and the other as a sensing arm. Since both the arms see different environmental conditions, even a slight change in ambient conditions can alter the output signal. This problem with the single-mode fiber optic interferometric sensors can be overcome by using dual-mode optical fiber sensors, where the reference and sensing arms of the interferometer are two modes of the same fiber [1–3]. Dual-mode fiber sensors are thus more stable than the single-mode fiber interferometric sensors. The purpose of the present experiment is to understand the working of a dual-mode fiber sensor. More specifically, we study a dual-mode fiber interferometer as a temperature sensor and measure its temperature sensitivity.

19.2 Aim

To study a dual-mode fiber interferometer as a temperature sensor and to measure its temperature sensitivity.

19.3 Apparatus

He-Ne laser, 20X microscope objectives, pinhole masked photodetector connected to a multimeter, power meter, thermometer, heating arrangement, around 2 meter long dual-mode fiber, and 2 XYZ-translational stages.

19.4 Theory

When both the modes of a dual-mode fiber are excited with almost equal power, one observes two intensity lobes at the output end of the fiber in the far-field pattern. It can be shown analytically that the intensity of both the lobes varies sinusoidally with the relative phase difference (ϕ) between the two modes at the output end of the fiber. Thus, the intensity distribution will be repeated after every phase change of 2π. The intensity evolution of the two lobes in the far-field pattern of a dual-mode fiber with ϕ is shown in Fig. 19.1. The phase difference ϕ can be changed due to some physical parameter(s) affecting the two modes differently. For example, if the temperature of the fiber is changed, the phase difference ϕ changes due to a change in the V-value and the length of the fiber, and hence, the intensity of both the lobes varies periodically. Thus, by knowing this intensity variation with temperature, one can measure the corresponding change in ϕ and the temperature. The sensitivity of such a sensor, defined as the change in relative phase difference ($\delta\phi$) per degree centigrade per unit length of the fiber, is given by

$$\text{Temperature sensitivity} = \frac{1}{L}\frac{\partial \delta\phi}{\partial T} = \frac{2\pi}{L\Delta T_{2\pi}} \qquad (19.1)$$

where, $\Delta T_{2\pi}$ represents the change in temperature responsible for $\delta\phi = 2\pi$ and L is the sensing length of the fiber. In order to demonstrate the working of such a sensor and to measure its temperature sensitivity, the following experiment is suggested.

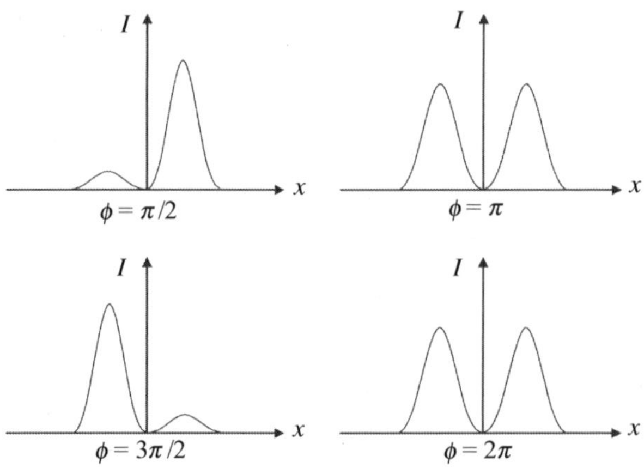

Fig. 19.1: Evolution of the far-field intensity pattern of a dual-mode fiber as the relative phase difference between the modes at the output changes. It is assumed that both the modes carry equal powers.

19.5 Procedure

The experimental set-up is illustrated in Fig. 19.2. To carry out the experiments, the following procedures have to be executed step by step:

1. Prepare the fiber ends so that the fiber has well-cleaved ends.
2. Mount the source in the aligner and adjust with the help of the aligning screws.
3. As discussed in Chapter 5 couple the light from a He-Ne laser source into the given fiber, which is dual-moded at the He-Ne laser wavelength.

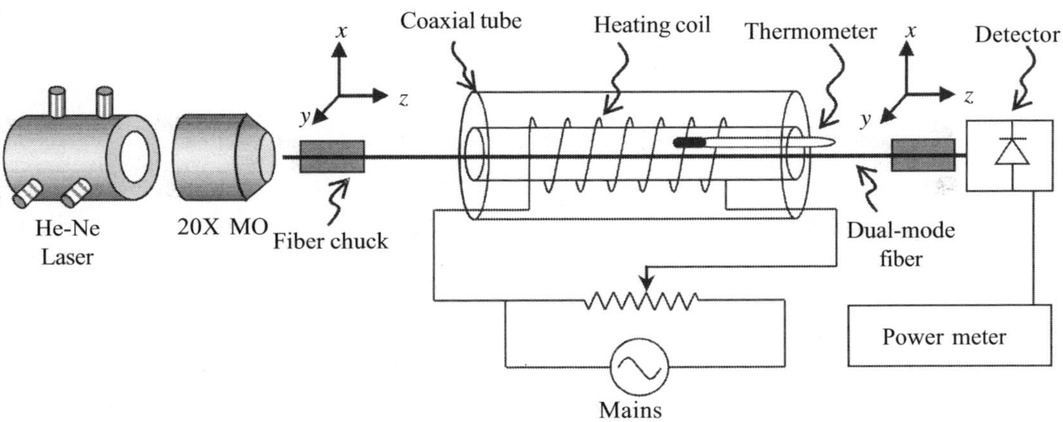

Fig. 19.2: Experimental set-up to illustrate the use of a dual-mode fiber as a temperature sensor.

4. Allow a section of the fiber of length L to pass through a heating arrangement consisting of a canthal heating wire of suitable length wrapped around a quartz tube to provide a uniform temperature around the fiber as shown in Fig. 19.2. The tube should be surrounded by another quartz tube to prevent heat loss to the atmosphere. The ambient temperature of the fiber can be changed by changing the voltage across the heating wire through a power divider arrangement.
5. Couple the light emerging from the output end of the fiber to a pinhole masked detector connected to a power meter. Adjust the input coupling in such a way that both the modes of the fiber get excited almost equally. This can be achieved by coupling light into the fiber at an off-axis point and by observing the variation in the intensity of either of the lobes in the far-field pattern as a function of temperature. When both modes are excited almost equally, the difference between maximum and minimum intensity of the lobe would be maximum. Adjust the position of the detector in such a way that the pinhole is positioned at the brightest point of one of the intensity lobes.
6. Note down the applied temperature to the section of the fiber by using a thermometer placed close to the fiber and the corresponding detector output.

19.6 Observations

Length of the fiber =

Least count of the thermometer =

Least count of the power meter =

S. No.	Applied temperature (°C)	Output power (mW)
1.		
2.		
.		
.		

19.7 Result

The intensity variation with temperature in a typical experiment is shown in Fig. 19.3. In this figure, circles show experimentally observed values. The sinusoidal variation of the intensity

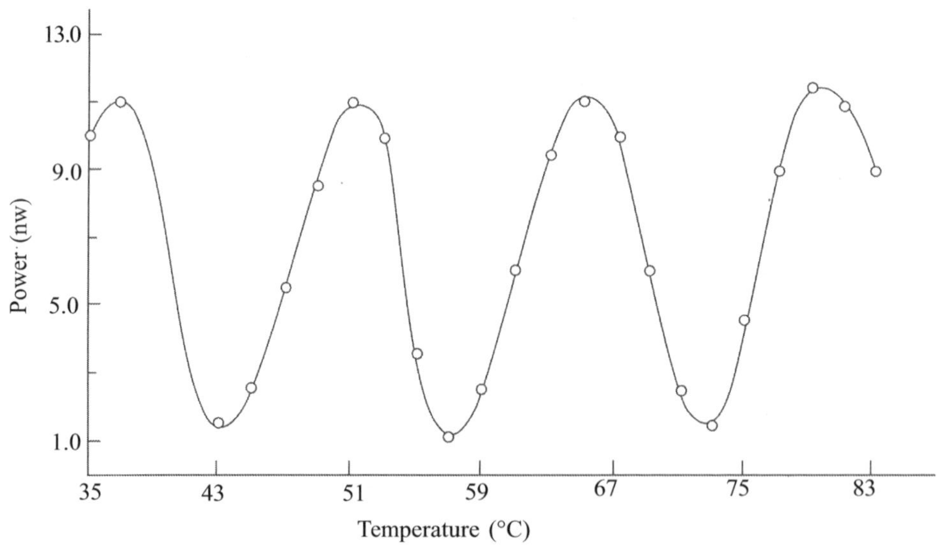

Fig. 19.3: Typical variation of power (in a lobe) with temperature.

with temperature illustrates the working of the sensor. From this figure one can easily calculate $\Delta T_{2\pi}$ which comes out to be ≈14°. In this particular experiment L was taken as 37 cm, and hence, the temperature sensitivity ($=2\pi/L\Delta T_{2\pi}$) was observed to be ≈1.2 radian/°C/meter.

References

1. M.R. Layton and J.A. Bucaro, "Optical fiber acoustic sensor utilizing mode-mode interferometer", *Applied Optics*, vol. 18, p. 666 (1979).
2. D. Kreit, R.C. Youngquist and D.E.N. Davies, "Two-mode fiber interferometer/amplitude modulator", *Applied Optics*, vol. 25, p. 4433 (1986).
3. K.A. Murphy, Mark S. Miller, A.M. Vengsarkar and R.O. Claus, "Elliptical-core two mode optical fiber sensor implementation methods", *J. Lightwave Technol.*, vol. 8, p. 1688 (1990).

20

Fiber Optic Mach-Zehnder Interferometer

K. Thyagarajan and Varghese Paulose

20.1 Introduction

One of the important applications of optical fibers is in the sensing of various external parameters such as pressure, temperature, magnetic field etc. A fiber optic Mach-Zehnder interferometer is one of the basic configurations used in high sensitivity measurements. In such an arrangement, changes in the phase of light propagating through a single-mode fiber caused due to an external parameter are very sensitively measured using an interference effect. Fiber optic Mach-Zehnder interferometers are one of the most sensitive sensors with predicted capabilities exceeding those of many conventional sensors. In this chapter, we will discuss the basic configuration and study how it can be used for sensitive measurements.

20.2 Aim

To set-up a Mach-Zehnder interferometer (MZI) with single-mode fibers for demonstration in the laboratory.

20.3 Apparatus

He-Ne Laser with holder, three XYZ-translation stages with V-grooves mounted on them, two 20X microscope objectives (MO), two MO holders, two cubic beam splitters, two prism tables, one V-groove, one magnifying glass, three posts of appropriate sizes, two half meter lengths of a single-mode fiber and a screen.

20.4 Theory

Figure 20.1 shows a conventional Mach-Zehnder interferometer using bulk optical elements in which light from a source is first split into two arms using a beam splitter and then recombined by another beam splitter after propagating through two different arms [1]. Depending on the phase difference between the two arms, at the output of the interferometer one obtains an interference pattern much like in any other interferometer. The corresponding fiber optic version of the Mach-Zehnder interferometer[1] is shown in Fig. 20.2 in which the two arms of the interferometer are now replaced by the two pieces of a single-mode fiber [2]. The beams emerging from the two fibers will almost be diverging spherical waves, which are superposed using the beam splitter. If the output ends of the two fibers are adjusted as shown in Fig. 20.3 then the two diverging spherical waves will have their centers of curvature collinear with the fiber axis. In such a case, circular fringes will be formed on a screen placed as shown in Fig. 20.2.

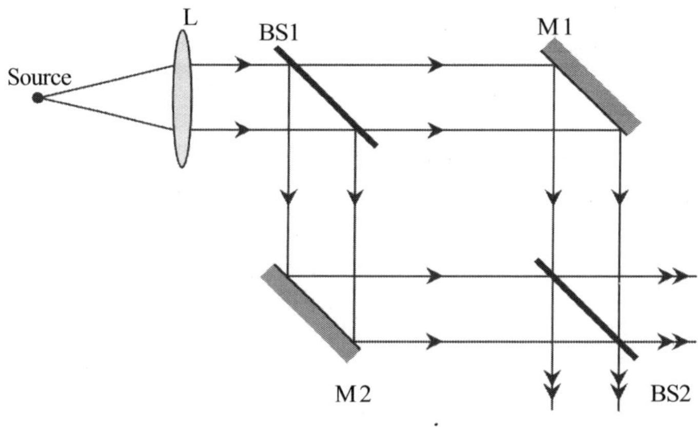

Fig. 20.1: Configuration of a conventional Mach-Zehnder interferometer (MZI).

This arrangement is very similar to the interference pattern formed by two point sources on a screen placed perpendicular to the line joining the sources. In contrast to a point source which emits uniformly along all directions, in the present case, the interfering beams have almost a Gaussian amplitude distribution. Thus, the circular fringe pattern will have an overall intensity pattern like that in a Gaussian beam. When the output ends of the fiber are brought closer to

[1] If the two beam splitters are replaced by fiber optic directional couplers, we would then have an all-fiber Mach-Zehnder interferometer.

each other, the circular fringes will expand and when they are moved away from each other they will contract (where is the zero order fringe in this system?). The total phase difference between the interfering beams at any point Q on the screen is given by (see Fig. 20.4)

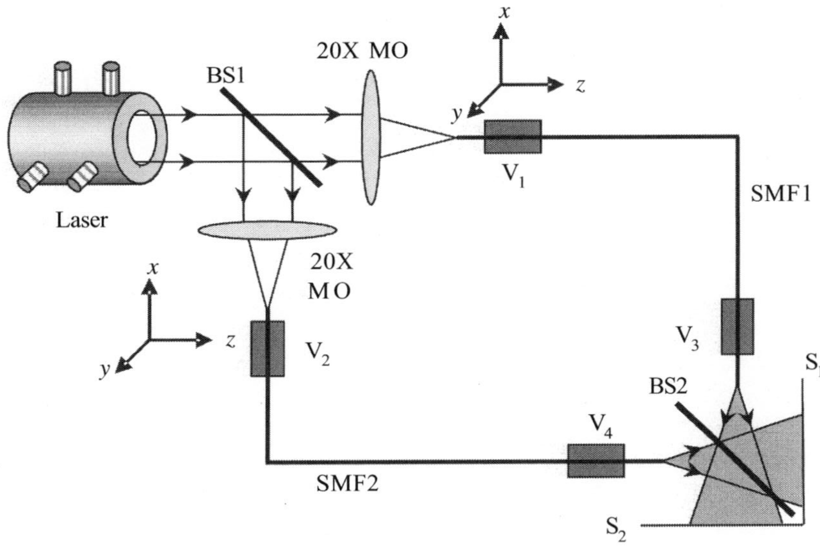

Fig. 20.2: Schematic of fiber optic MZI.

$$\begin{aligned}\phi &= \frac{2\pi}{\lambda_0}(S_2Q - S_1Q) + (\phi_2 - \phi_1) \\ &= \frac{2\pi}{\lambda_0}\left(\sqrt{R_2^2 + r^2} - \sqrt{R_1^2 + r^2}\right) + (\phi_2 - \phi_1)\end{aligned} \quad (20.1)$$

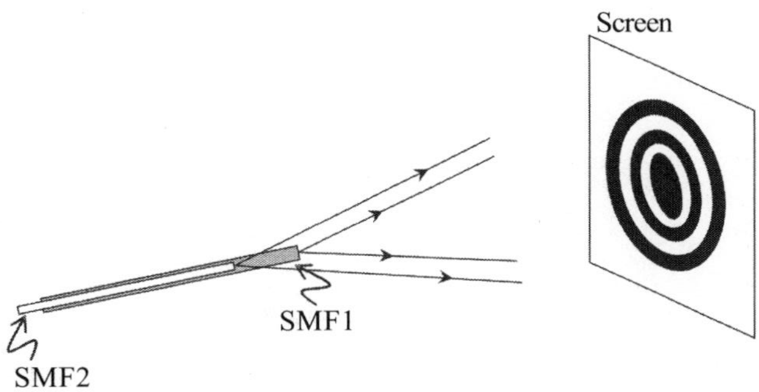

Fig. 20.3: The effective fiber configuration due to the beam splitter BS2 at the output, which results in circular fringes. Fringes can be made to collapse or expand by moving one fiber towards or away from the beam splitter.

where r is the distance of the observation point from the center of the screen, R_1 and R_2 are the distances of the fiber ends from the screen. The quantition ϕ_1 and ϕ_2 represent the phase shifts suffered by the beam in propagating through the fiber; these include any constant phase difference between the two arms due to differences in fiber lengths etc. as well as any difference due to external perturbation induced phase change.

If each of the beam splitters is a 50 : 50 beam splitter, then the averaged intensity detected by an optical detector would be

$$I = I_0 \cos^2\left(\frac{\phi}{2}\right) \tag{20.2}$$

where I_0 is the input intensity. It is seen from Eq. (20.2) that the intensity at any point depends on the phase difference between the two arms. Thus, if the phase difference ϕ is an integral multiple of 2π, then the intensity is maximum while if it is an odd integral multiple of π, then the intensity would be zero (what happened to energy conservation?). Thus, the intensity at any point on the screen depends on the phase difference between the two arms of the interferometer. Any external perturbation differentially affecting the light propagating through the two arms would change $(\phi_2 - \phi_1)$ and thus the interference pattern itself.

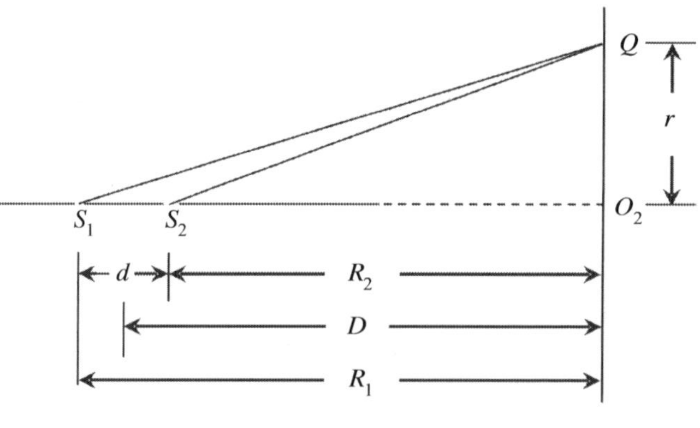

Fig. 20.4: The ray diagram for interference between two point sources. The screen is perpendicular to the line joining the two sources.

A fiber optic Mach-Zehnder interferometer as shown in Fig. 20.2 can be used as a sensor for detecting and measuring various external parameters such as temperature, pressure, strain, magnetic field etc. Thus, if only one of the arms of the Mach-Zehnder interferometer (called the signal or the sensing arm) is subjected to temperature change or to an acoustic wave then the beam propagating through the sensing arm will suffer a relative phase change with respect to the other arm (referred to as the reference arm) due to a change in the refractive index and physical length of the fiber by these measurements. This change in the phase can then be detected as an intensity change in the output of the interferometer. If n represents the refractive index of the two fibers and l their lengths then the change of phase between the two arms due to an external effect is given by

$$\Delta\phi = \frac{2\pi}{\lambda_0}\Delta(nl) \tag{20.3}$$

where λ_0 is the wavelength of light in free space. For the output intensity to change from a maximum to a minimum,

$$\Delta\phi = \pi \tag{20.4}$$

Assuming $n = 1.5$, $\lambda_0 = 1\ \mu m$ and $l = 1$ m, we obtain the corresponding fractional change in path length to be

$$\frac{\Delta(nl)}{nl} = 3 \times 10^{-7} \tag{20.5}$$

Today's limit of laboratory technology with very demanding signal processing can detect phase changes of 10^{-7} radians. Since a fractional change of one part in 3×10^6 leads to a change in phase of π, one can indeed detect fractional changes of optical path lengths as small as 10^{-13} radians. Indeed, since the interferometer is extremely sensitive, we can even detect changes in the output intensity by bringing our hand close to one of the fibers and hence changing its temperature even so slightly.

Since this is an interference experiment, for a proper interference pattern, the two beams emerging from the fibers should have the same state of polarization (SOP). Since conventional single-mode fibers do not, in general, maintain the SOP, placing a polarizer at the output (after the output beam splitter) can sometimes give better contrast. To get best contrast one could use polarization maintaining fibers with their axes properly aligned.

20.5 Procedure

The experimental set-up is illustrated in Fig. 20.2. To carry out the experiments, the following procedures have to be executed step by step:

1. Align the laser with the beam splitter (BS1) and the two MOs so that all the reflections, from the two MOs through the beam splitter, go back (very nearly) into the laser.
2. Fix the fibers in the V-grooves V_1 and V_2 mounted on the XYZ-stages, and align for maximum coupling from the laser. Try to couple as much light as possible in the two fibers individually, and take care to see that the magnitude of the light coupled in both the fibers are nearly the same.
3. Place the output end of one fiber on the V-groove (V_3) of the third XYZ-stage and the output of the other fiber on a separate V-groove (V_4), at right angles to V_3 so that both fiber ends are nearly equidistant from the beam-splitter BS2, as shown in Fig. 20.2.

4. Adjust the positions of V_3, V_4 and BS2 (which is placed on a prism-table) so that the two beams coming out of the two fiber-ends overlap on the screen, after passing through BS2.
5. Now, keep the magnifying glass (MG) at the output of BS2 so that the two light beams pass through it and are focused on the screen.
6. Adjust the position of V_3 so that the focused spots coincide.
7. Remove the magnifying glass to observe circular fringes on the screen.
8. Again, adjust the position of the fiber-end placed on V_3 using the longitudinal movement to see the circular fringes clearly. Using this movement, you can increase or decrease the diameter of the central fringe.

The fiber optic Mach-Zehnder interferometer is now ready. Observe the continuous movements of fringes back and forth (i.e. collapsing towards the center, and originating from the center). Try to explain your observations. Remember that interferometers are very sensitive to phase changes! If you bring your hand close to one of the fibers, you can observe rapid collapse (or expansion) of the fringes. This is caused due to the minute change of the temperature of the fiber resulting in a corresponding refractive index change. Try bringing your hand close to the other fiber arm – what do you observe?

20.6 Comments

To have a deeper insight and understanding, one can further carry out the following experiments:

1. Instead of using a beam splitter at the output end, you may try putting the two fiber ends (with the plastic jacket removed) touching each other and lying parallel as shown in Fig. 20.5. You would then observe straight line fringes on the screen. By measuring fringe spacing and the distance of the screen from the fiber end, estimate the core diameter.

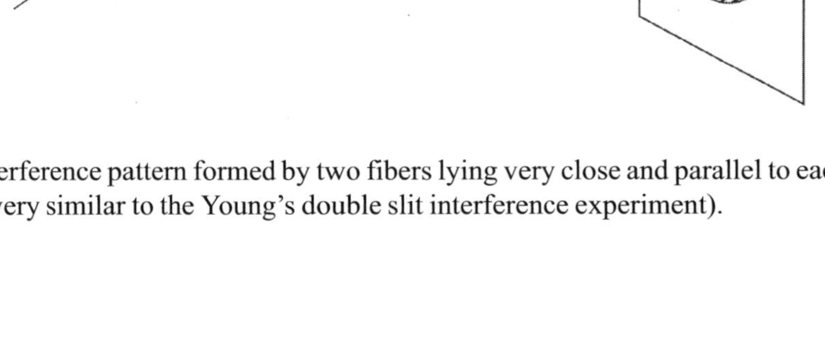

Fig. 20.5: Interference pattern formed by two fibers lying very close and parallel to each other. (This is very similar to the Young's double slit interference experiment).

2. One can use the set-up for a simple measurement of the wavelength of laser. From Eq. (20.1) we can show that the order of the bright fringe of radius R is

$$m_R = \frac{d}{\lambda_0}\left(1 - \frac{R^2}{2D^2}\right) + \frac{\phi_2 - \phi_1}{2\pi} \tag{20.6}$$

where D and d are defined in Fig. 20.4, and we have assumed that $D \approx R_1 \approx R_2 >> d$. The order of the central fringe at O_2 is

$$m_0 = \frac{d}{\lambda_0} + \frac{\phi_2 - \phi_1}{2\pi} \tag{20.7}$$

Hence the number of bright fringes contained within a circular radius of R will be

$$N = m_0 - m_R = \frac{R^2 d}{2D^2\lambda_0} = \frac{R^2}{2D^2\lambda_0}(d_0 + \Delta) \tag{20.8}$$

where d_0 is some chosen arbitrary separation between the fiber ends, and Δ the displacement from this position. In order to calculate λ_0, you should measure N, the number of bright fringes within a chosen circle of radius R as a function Δ; Δ is changed by moving one of the fiber ends by a known amount. A plot of N vs. Δ would be a straight line whose slope would be $R^2/2D^2\lambda_0$. Knowing R and D, you can calculate λ_0.

One of the attractive features of the above method is that even if the fringes are not steady, one can measure N within one fringe and thus be able to estimate λ_0.

Finally, while carrying out this experiment, the following points should be taken care of:

1. Do not look into the laser or into the output of the fiber (It may burn a spot on the retina).
2. Difference in optical path length of the fibers should be less than the coherence length of the laser, otherwise no fringes will appear. Typical coherence lengths are about 10 cm and since the refractive index of the fiber is about 1.5, the physical length difference should be less than 6 cm or so.
3. Do not touch the fiber input or output ends with your hands because this might lead to dust, grease, oil or any other such material settling on the cleaved surface of the fiber ends, thus affecting the coupling efficiency.

Reference

1. M. Born and E. Wolf, *Principles of Optics*, 6th Edition, Pergamon Press, Oxford (1980).
2. A.K. Ghatak, *Optics*, 4th Edition, McGraw-Hill India, New Delhi (2008).

21 Optical and Electrical Characteristics of a Light Emitting Diode

M.R. Shenoy

21.1 Introduction

In its simplest form, a light emitting diode (LED) consists of a *p*-type doped semiconductor and an *n*-type doped semiconductor forming a *p-n* junction. When the LED is forward biased, charge carriers flow across the junction and it is the recombination of the positive (holes) and negative (electrons) charge carriers that results in the emission of light from the vicinity of the *p–n* junction. One of the major applications of LEDs is in electronic displays. LEDs are also used as a source in optical fiber communication. A simple optical fiber communication system consists of a transmitter in the form of an LED whose output is modulated by the information-bearing signal and this modulated output (optical signal) is transmitted via an optical fiber. The system receiver consists of an avalanche photodiode (APD) or a PIN diode. It converts the optical signal into the corresponding electrical signal, which on demodulation gives the original information-bearing signal. Due to its small size, low power consumption and the requirement of only a simple driving circuit, LEDs form a very useful source in fiber optic communications, particularly for short-range applications. LEDs used in optical communication usually radiate in the near infrared region of the electromagnetic spectrum, and it is in this wavelength range that the transmission properties of optical fibers are best utilized.

Before using an LED as an optical source in a fiber communication system, it is important to know the basic optical and electrical characteristics of the LED, such as: (a) *V-I* characteristics, (b) Maximum forward current, (c) Radiant power, (d) Peak emission wavelength, (e) Spectral bandwidth, (f) Radiation pattern/directionality, (g) Frequency response/modulation efficiency, (h) Switching characteristics/rise time/fall time, etc.

In the following sections, we describe simple procedures to estimate some of these characteristics of an LED. For details about the construction, materials and operation of an LED, the reader may refer to the references listed at the end of this chapter.

21.2 Aim

To study the optical and electrical characteristics of a given LED.

21.3 Apparatus

Given LED, power supply, multimeter, monochromator, rotation stage, detector, power meter, function generator, CRO, etc.

21.4 Procedure

21.4A. *V-I* Characteristics

The electrical circuit diagram is shown in Fig. 21.1(a). V_S is a variable voltage source and R_S is a series resistor that limits the maximum current through the LED. Since there is a maximum limit on the forward current *I* for safe operation of an LED (≈100 mA), R_S has to be chosen according to the relation

$$R_S = \frac{(V_S)_{max}}{(I_F)_{max}} \tag{21.1}$$

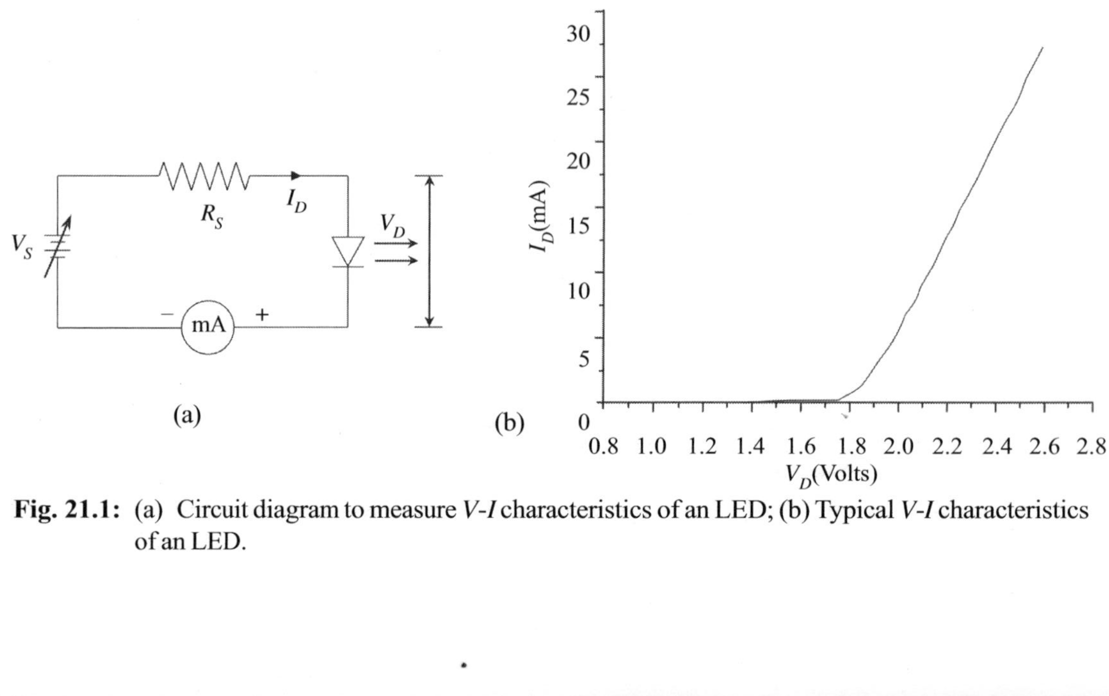

Fig. 21.1: (a) Circuit diagram to measure *V-I* characteristics of an LED; (b) Typical *V-I* characteristics of an LED.

In writing Eq. (21.1), we have neglected the forward resistance of the LED, which is typically a few ohms. Usually the manufacturer specifies the value of $(I_F)_{max}$. A rough estimation of $(I_F)_{max}$ may be made by feeling the temperature of the LED by finger touch when a forward current I_F flows through the diode. If the LED becomes very hot (say above 70°C) when the forward current I_F is increased to a certain value, it implies that I_F is close to $(I_F)_{max}$. Care should also be taken to see that the power dissipation rating of the resistor R_S is greater than $(I_F)^2_{max} R_S$.

In order to determine the current through the diode as a function of the applied voltage, a milliammeter may be connected in series to directly read I_D. On the other hand, we may measure accurately the voltages V_S and V_D using a digital multimeter and obtain I_D as

$$I_D = \frac{V_S - V_D}{R_S} \tag{21.2}$$

The experiment involves changing V_S from zero to $(V_S)_{max}$ in suitable steps and measuring (or calculating) I_D. A plot of I_D vs. V_D represents the *V-I* characteristics (see Fig. 21.1b). The *V-I* characteristics enable the user to choose an operating point Q in the linear region of the characteristic curve. This is essential in order to avoid any distortion that may be introduced during modulation of the LED output. In any circuit using an LED, the bias voltage is adjusted or chosen such that V corresponds to the required operating point Q.

While using an LED as a source of light, the diode is forward biased and therefore it is sufficient to determine the *V-I* characteristics in the forward biased mode. However, one can obtain the reverse *V-I* characteristics by simply interchanging the diode terminals and repeating the experiment. It is easy to estimate the breakdown voltage as well.

21.4B. *I-P* characteristics

This part of the experiment may be clubbed with the previous part, namely the *V-I* characteristics. Here what we are interested in is to know the relationship between the radiant optical power P and the current through the diode, I_D; in other words, to determine the variation of the optical power as a function of the current I_D. The circuit diagram remains the same as before. In addition, a large area photodetector is placed close to the LED (to collect almost all the radiant energy), and the photodetector is connected to a power meter. The LED and the detector have to be completely covered with a black cloth (unlike in the previous case) to avoid any ambient light being detected. Alternatively, the experiment may be done in a dark room.

Now the current I_D is increased from zero to $(I_F)_{max}$ in suitable steps by varying the biasing supply V_S, as before. The corresponding reading P on the power meter is noted at each step. A graph is plotted between I_D and P, which represents the *I-P* characteristics (see Fig. 21.2).

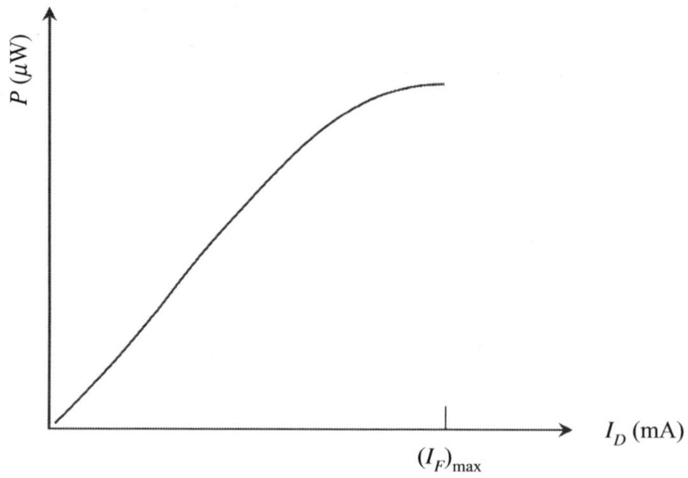

Fig. 21.2: Typical *I-P* characteristics of an LED.

From the *I-P* characteristics it may be seen that the optical power radiated by an LED is almost linearly proportional to the current I_D, except for the large values of I_D when it approaches $(I_F)_{max}$ (Why?). From the *V-I* and *I-P* characteristics, we reach to the following important conclusion: If the LED is biased at the *Q* point in the linear region of the *V-I* characteristics, then for small signals, the output optical power of the LED is proportional to the input signal. In other words, there will be no (or very little) distortion in the modulated optical signal.

21.4C. Modulation bandwidth

Since the output power of an LED is linearly proportional to the input current (I_D), intensity modulation is possible by adding a signal current to the bias current so that the optical current, and therefore the optical power can vary linearly over a suitable range. However, there are limitations on the maximum frequency at which the LED output can be modulated depending on the material and structure of the device. The small signal modulation bandwidth of an LED may be defined as the frequency at which the modulation efficiency falls to one half of its maximum value (at low frequencies). The modulation efficiency is given by

$$\eta = \frac{P_{max} - P_{min}}{P_{max} + P_{min}} \tag{21.3}$$

where P_{max} and P_{min} are the output optical powers corresponding to V_{max} and V_{min} of the modulating signal. Alternatively, if we can directly measure the modulated a.c. (output) signal corresponding to the modulating (input) signal, then the modulation bandwidth corresponds to the (high) frequency at which the amplitude of the detected a.c. signal falls to one half of its value at low frequencies. A schematic of the experimental set-up is shown in Fig. 21.3(a).

In order to determine the modulation bandwidth of an LED, the LED is biased at the operating point and a simple sinusoidal output from an oscillator is superimposed on the DC bias voltage. One can achieve this simply by replacing the DC power supply V_S by an oscillator which is

(a)

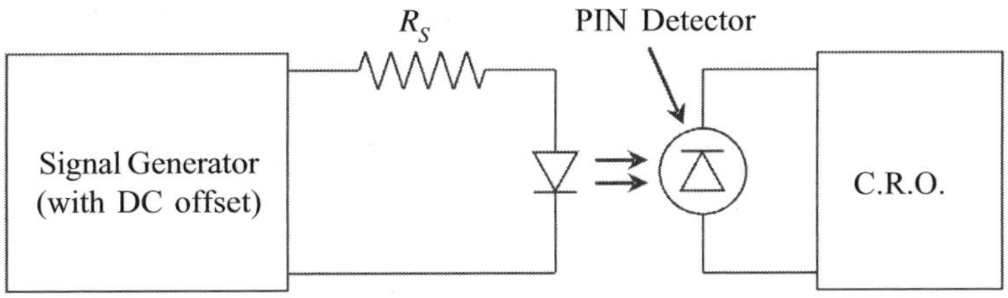

(b)

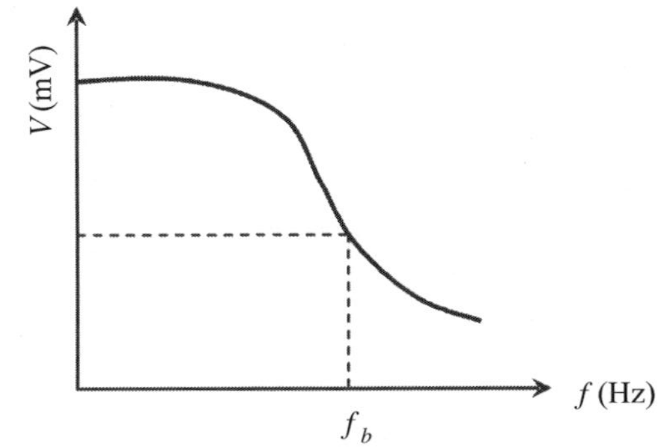

Fig. 21.3: (a) Schematic of an experimental set-up to measure modulation bandwidth of an LED; (b) Typical frequency response curve of an LED.

provided with a DC offset for the output. Keeping the oscillator output to a minimum, the DC offset control is adjusted such that V_D, the d.c. voltage across the LED, corresponds to the Q-point. A PIN detector, placed in front of the LED, is connected to an oscilloscope. The LED output is now modulated by applying small-amplitude (Why?) low-frequency sinusoidal signal (e.g., 0.2 V peak-to-peak) from the signal generator. The display on the CRO shows a corresponding sinusoidal signal. The detected signal peak-to-peak voltage is measured. The frequency of the oscillator output is increased in suitable steps keeping the amplitude constant, and the corresponding amplitude reading on the oscilloscope is recorded till the amplitude falls well below one half of its initial value (at low frequency). A graph of frequency vs. output signal amplitude gives the frequency response curve [see Fig. 21.3(b)]. The modulation bandwidth is given by the frequency at which the amplitude falls to one half of its initial constant value. Since optical sources operate down to d.c., we only consider the high frequency 3 dB point.

It should be mentioned that by following the above procedure, what has been estimated is the optical 3 dB bandwidth. The electrical 3 dB bandwidth corresponds to the frequency at which the amplitude of the detected signal falls to 0.707 times its initial constant value (Why?). Ordinary display LEDs typically have a modulation bandwidth of a few tens of KHz, while communication or switching grade LEDs may have a bandwidth of several tens of MHz or more. It may be pointed out that the modulation bandwidth of an LED is not independent of the biasing point (even in the linear region of the *V-I* characteristics). It is worthwhile to see that the modulation bandwidth increases with increase in the d.c. bias voltage.

21.4D. LED output spectrum

Every light source, in general, has a certain characteristic wavelength spectrum and radiation pattern (angular spectrum) of its output radiation. Since the propagation characteristics of an optical fiber are wavelength dependent, it is very important to know the wavelength spectrum of the LED output. The radiation pattern, which gives a measure of the directionality of the emitted radiation, determines the launching efficiency of light into the fiber when used as a source in optical fiber systems.

i) Wavelength spectrum

A schematic experimental set-up to measure the wavelength spectrum of an LED is shown in Fig. 21.4(a). The supply voltage V_S is adjusted to bias the LED at the *Q*-point (actually, in this particular case, it is preferable to bias the LED at a slightly higher voltage in order to obtain a higher optical output). The input slit of the monochromator, placed close to the LED, is aligned with the LED's output direction. The PIN detector, attached to an optical power meter, is placed close to the output slit of the monochromator. There should be no (or minimum) ambient light in the vicinity of the set-up. (In fact, it is preferable to perform this experiment in a dark room.) The transmission wavelength of the monochromator is now varied in suitable steps over an appropriate range of wavelengths, and at every step the corresponding power meter reading is recorded. A graph is plotted between wavelength and output power that represents the wavelength spectrum of the LED [see Fig. 21.4(b)].

From the graph, the peak emission wavelength (which corresponds to the peak of the spectrum) and the bandwidth or 'linewidth' $\Delta\lambda$ (which is the full-width at half maximum) are obtained. While communication grade LEDs have typical linewidths of about 30 nm or less, display LEDs may have linewidths in excess of 100 nm, since linewidth is not a critical parameter in the latter case (Why?).

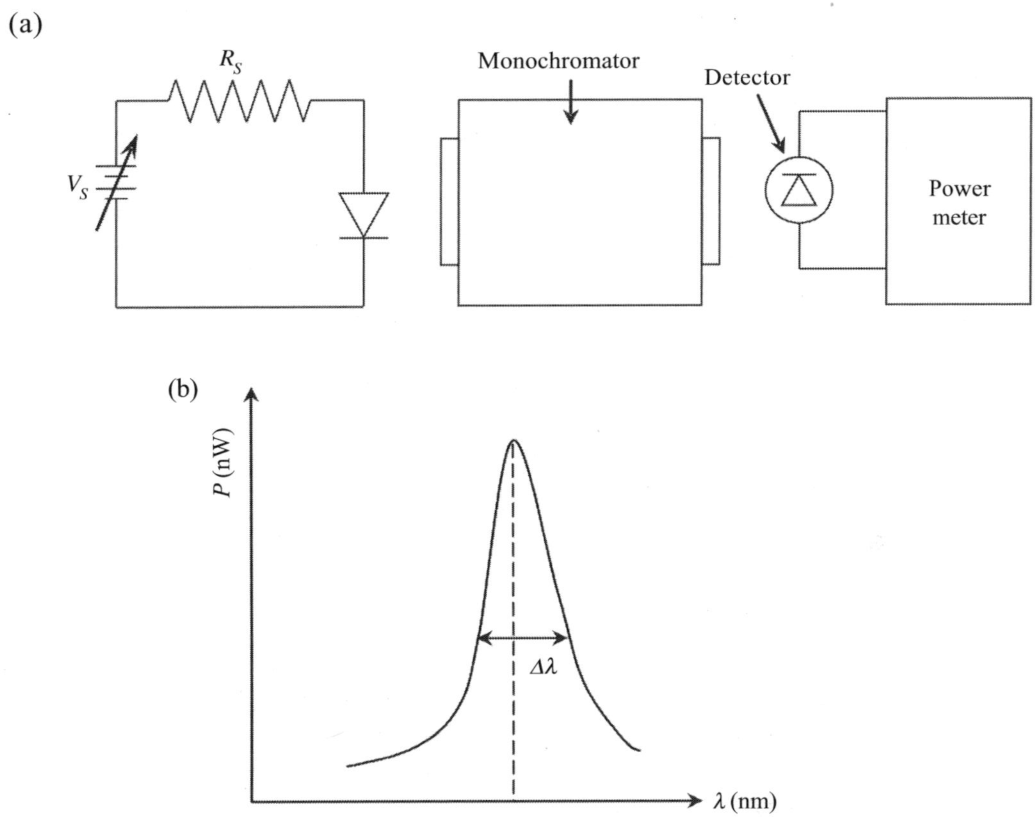

Fig. 21.4: (a) Schematic of the experimental set-up to measure the wavelength spectrum of an LED; (b) Typical wavelength spectrum of an LED.

ii) Radiation pattern

To obtain the radiation pattern of an LED one has to scan the LED output in the angular direction in a plane intersecting the axis of the output cone of radiation. A simple method to do this is to place the LED appropriately at the axis of a spectrometer table and scan the output using a pinhole detector fixed on to the telescope arm [see Fig. 21.5(a)].

The pinhole (in front of the detector) should lie in the scanning plane (which in this case is a horizontal plane) and at the same level as the LED. Care should be taken to center the LED on the axis of rotation. The pinhole detector is connected to a power meter. Since a pinhole detector is to be used, the distance between the detector and the LED should not be large in order to make the detected power easily measurable. However, keeping the detector close to the LED would affect the angular resolution. (In fact, one should estimate the optimum distance for a given size of the pinhole and the angular resolution of the spectrometer table.) Now the detector arm is brought to a position perpendicular to the output direction. The power meter reading $P(\theta)$ is recorded at every step as the detector arm scans the LED output through 180° (or more) in suitable steps. A polar graph is drawn between the output power and angle. Usually, one draws the relative power P_r $\left(= \dfrac{P(\theta)}{P_{max}}\right.$, where P_{max} is the maximum power recorded corresponding to an angle $\theta = \theta_{max}$) against the translated angle θ_t $(= \theta - \theta_{max})$ to obtain the 'lobe' structure [see Fig. 21.5(b)].

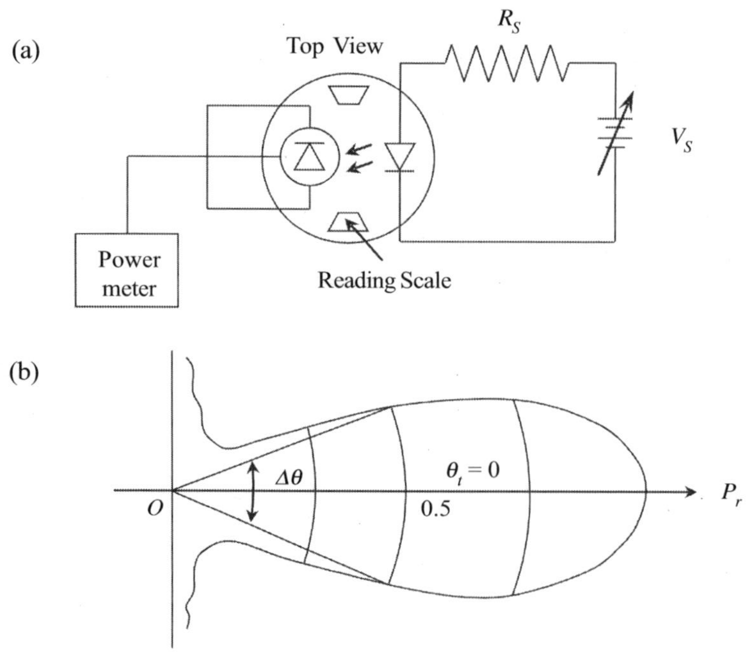

Fig. 21.5: (a) Schematic of the experimental set-up to measure the radiation pattern of an LED; (b) Typical polar diagram of radiation pattern of an LED.

21.5 Comments

1. In the entire procedure described above it is assumed that the detector response is linear over the whole range of measured power, and that the bandwidth of the detector is sufficiently high. Therefore one must be sure that the detector used, satisfies these conditions. In fact, one can outline a similar experiment to characterize a given detector.
2. The plots of the various characteristics illustrated in this chapter are typical of red (display type) LEDs. It should be pointed out that actual communication grade LEDs have much better characteristics in terms of smaller linewidth, better directionality, larger output power, higher modulation bandwidth, etc.
3. The speed of response of optical sources is largely dictated by their respective radiation emission mechanisms. The response of an optical source to a (current) step input is often specified in terms of the 10–90% rise time, a parameter which is reciprocally related to the device frequency response or the modulation bandwidth. It must be mentioned, however, that the modulation bandwidth of a source depends also on the driving circuit, and can be maximized using suitable drive circuits. The procedure described in this chapter to estimate the modulation bandwidth is aimed at qualitative demonstration of the concept.
4. The parameters, characterizing a semiconductor laser diode (LD) are also the same as those characterizing an LED except that there exists a "threshold value" of forward current above which the "lasing action" takes place and the output power increases rapidly with the forward current. Usually, LDs emit radiation in the near IR (invisible) region and therefore, care should be taken to avoid direct/reflected radiation entering the eyes.

References

1. Y. Suematsu and K. Iga, *Introduction to Optical Fiber Communications*, John Wiley and Sons (1982).
2. John M. Senior, *Optical Fiber Communications: Principles and Practice*, Prentice Hall International (1985).
3. W.B. Jones, Jr., *Introduction to Optical Fiber Communication Systems*, Holt-Rinehart-Winston (1988).
4. H. Kressel (Ed.), *Semiconductor Devices for Optical Communication*, Topics in Applied Physics, vol. 39, Springer-Verlag, Chapter 2 (1980).
5. L. Sharupich and N. Tugov, Opto-electronics, MIR Publishers (1987).
6. S.E. Miller and I.P. Kaminow (Eds.), *Optical Fiber Telecommunications II*, Academic Press Inc. (1988).

22

Characterization of Fiber Bragg Grating - I

Sunil K. Khijwania and C.S. Goh

22.1 Introduction

Fiber Bragg gratings (FBGs), with intra-core periodic/a-periodic refractive index (RI) perturbation along the fiber length, have emerged as a key passive component with a widespread applications including but not limited to fiber laser, laser wavelength stabilization, WDM demultiplexer and add/drop filter, optical amplifier gain equalization and dispersion compensation in long-haul optical communication systems along with sensing applications. Main advantages of FBGs include compact all-fiber configuration, low insertion loss, high return loss or extinction, ease of coupling (with other fibers), and polarization insensitivity. However, the most distinct feature of FBG is the flexibility to achieve desired spectral characteristics as per the need of a given application, because it's the spectral characteristics of FBG, which play a significant role while employing it in an all-optical communication network or to realize an all-optical sensor system. Hence, one needs to incorporate FBG with proper optical characteristics into the optical network for such applications. This chapter introduces the basic concepts of FBG and their optical characteristics. Further, a simple experimental technique is introduced to have elementary estimation of these characteristics of FBG. Concepts of apodized and aperiodic gratings along with a more advanced and accurate technique to characterize FBG for its applications in optical communication systems will be discussed in Chapter 23.

22.2 Aim

To study the optical spectrum of a given fiber Bragg grating and determine its basic characteristics.

22.3 Apparatus

Broadband light source (ASE/white light), optical spectrum analyzer (OSA), optical circulator/3 dB coupler, isolator, optical fiber splicer, patch cords, fiber connector cleaner.

22.4 Theory

22.4.1 Fiber Bragg Grating

Exposure of a silica fiber core, which is doped with a material like germanium[1], to an intense laser radiation at certain wavelengths in ultraviolet (UV) spectral region induces a permanent change in its refractive index (RI). This phenomenon is termed as *photosensitivity* and was first observed by Hill and coworkers [1] in 1978 quite by an accident. While carrying out an experiment to investigate nonlinear effects in a highly germanium-doped silica core fiber with intense Argon-ion laser radiation at 488 nm, they observed a significant reduction in the transmitted light intensity over a prolonged laser exposure. During the same exposure interval, back-reflected (counter propagating) light intensity was observed to grow. The fiber essentially functioned like a reflection filter. This was attributed to the existence of a permanent RI modulation throughout the length of the fiber core. The forward propagating launched laser beam interfered with the backward propagating 4% Fresnel reflected beam from the far end of the fiber. This interference formed a standing wave intensity pattern throughout the length of the fiber. Absorption of the laser light by the germanium-doped silica in the bright (high-energy) regions resulted in a permanent increase of the local RI. Such fiber with permanent RI modulation within its core is termed as Fiber Bragg grating (owing to the reasons explained in Section 22.4.2). The RI change in such gratings was observed to be very small. As a result, grating strength was very poor. To achieve ~100% reflection, a very long length of grating was required. Lam et al. [2] showed that the grating strength increased as a function of the square of the light intensity. This suggested a two-photon process as the possible mechanism of grating formation. Thereby, a single photon at one half of this wavelength (at 244 nm in UV) was found to be more effective for the grating inscription. Microscopic reason attributed to the RI change is the formation of defect centers within the germania doped silica fiber core with an absorption peak at ~242 nm. UV irradiation induces photochemical reaction, which in turn yields modification of glass structure and hence changes absorption spectrum. This change in absorption leads to refractive index change [3]. The number of defect centers increases as a function of germanium concentration, which in turn, increases UV induced RI change (δn). As an example, $\delta n \sim 3 \times 10^{-5}$ for 3 mol% GeO_2 (standard telecommunication fiber), whereas $\delta n \sim 2.5 \times 10^{-4}$ for ~20 mol% GeO_2 doped fiber [4]. Photosensitivity of germania doped fiber can be further enhanced through the process

[1] To achieve step-index profile at core-cladding interface, silica fiber core is often doped with germania to increase core refractive index with respect to the cladding refractive index.

of *hydrogen loading*. For this, fiber is soaked in hydrogen gas under high pressure (typically 150 atm) in 20–75°C temperature range. δn as high as 10^{-2} has been realized by this method [3].

In 1989, Meltz et al. [5] demonstrated the interferometric/holographic fabrication technique for FBG inscription. Unlike Hill et al. [1], here the incoming UV beam was divided into two beams and then allowed to recombine at a pre-decided angle to form an interference pattern. Further, the photosensitive fiber was exposed from the transverse direction to this interference pattern (Fig. 22.1). At the location of bright fringe, RI increased, whereas at the position of dark fringe it remained unchanged. As the recombination angle between two beams along with UV wavelength decides fringe spacing and hence grating period, this technique opened up the possibility of realizing FBG at any characteristic operating wavelength. However, this technique is highly susceptible to vibration and requires stable optical set-up with sturdy base along with a highly

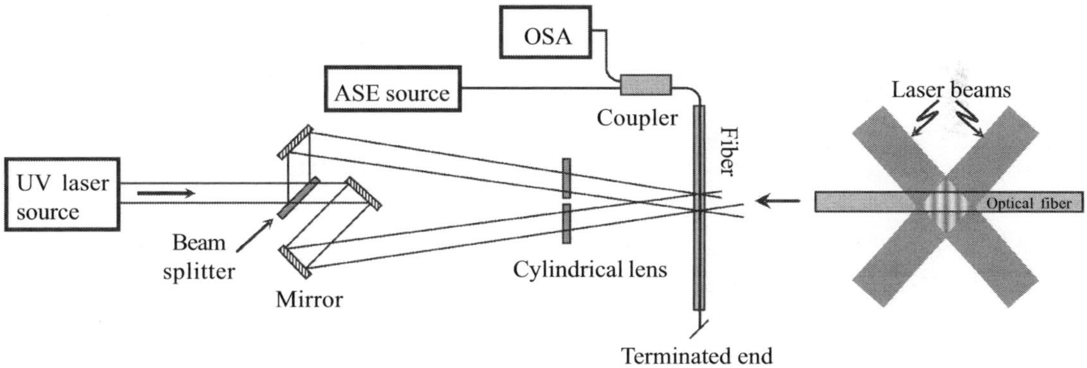

Fig. 22.1: A schematic of the fiber Bragg grating fabrication using holographic technique.

coherent UV source. Furthermore, the grating length is limited by laser beam spot-size. In 1993, Hill et al. [6] introduced a robust FBG fabrication technique based on the phase mask method, which overcomes these limitations and has been the most popular technique even today. A phase mask is a silica plate where surface relief grating of required period is etched at one

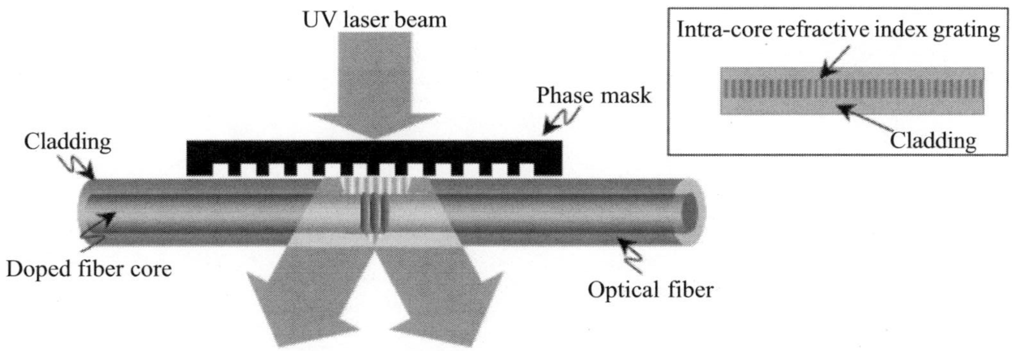

Fig. 22.2: A schematics of the fiber Bragg grating fabrication using phase mask method. Inset shows the resulted FBG in the overlap region of ±1 order diffracted radiation.

surface[2]. This periodic surface relief grating is optimized to suppress zero order diffraction and maximize ±1 orders [3]. In this method, normal incidence of a UV beam on a phase mask gives rise to an interference pattern of +1 and –1 diffracted orders. If a photosynthesized optical fiber is placed adjacent to the phase mask, intensity variation within the interference pattern gets imprinted in the form of a refractive index variation within the fiber core as shown in Fig. 22.2.

22.4.2 Fiber Bragg Grating as a guided-wave component

A fiber grating structure is a periodic modulation of refractive index along the fiber core. This structure can be designed to lead a coupling between (a) a forward propagating guided mode to a backward propagating guided mode (such gratings are termed as *short-period* gratings or FBGs), and (b) a guided fundamental mode to the co-propagating cladding modes (such gratings are termed as *long-period* gratings). We shall consider FBGs here. When light from a broadband source is launched into the fiber core carrying FBG, it gets scattered by each grating plane and suffers extremely weak reflection. However, under specific condition termed as Bragg condition, reflected light from each grating plane would be in phase and interfere constructively in the backward direction. This gives rise to a strong reflected spectrum centered at a specific wavelength termed as the *Bragg wavelength* (λ_B) (Fig. 22.3). On the other hand, other wavelength components would be transmitted by the grating, giving rise to the transmission spectrum of the FBG as shown in Fig. 22.3 (refer to Fig. 22.6 as well). Bragg condition is the manifestation of energy and momentum conservation, which requires the frequency of the coupled radiation (forward and backward) to be the same, and forward mode propagation constant $(\vec{\beta}_f)$ plus the grating wave-vector $(\vec{K})$ to be equal to the backward mode propagation constant $(\vec{\beta}_b)$. The second requirement can be stated as

$$\vec{\beta}_f + \vec{K} = \vec{\beta}_b \tag{22.1}$$

As $\vec{\beta}_b = -\vec{\beta}_f$ and $K = \dfrac{2\pi}{\Lambda}$ above condition simplifies to

$$2\left(\frac{2\pi}{\lambda_B}\right)n_{\text{eff}} = \frac{2\pi}{\Lambda} \tag{22.2}$$

or

$$\lambda_B = 2\Lambda n_{\text{eff}} \tag{22.3}$$

where Λ is the grating period (distance between two adjacent maximum points of the periodic refractive index), n_{eff} is the mode effective index and λ_B is termed as the *Bragg wavelength*

---

2 The phase mask is essentially a binary phase grating produced through controlled etching of a fused silica or quartz substrate by a computer-controlled e-beam writing technique similar to that used in forming amplitude masks in standard microelectronic manufacturing technology. The main difference here is that the mask after fabrication is transparent and it is a *phase* mask rather than an amplitude mask.

(see Fig. 22.3). For a silica fiber $n_{eff} \approx 1.46$, hence for $\lambda = 1550$ nm to be reflected, one must realize a grating with

$$\Lambda = \frac{\lambda_B}{2n_{eff}} = \frac{1550 \text{ nm}}{2 \times 1.46} \approx 0.531\ \mu m$$

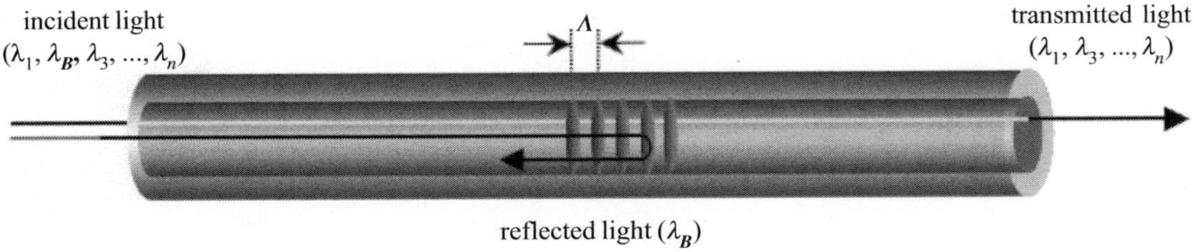

Fig. 22.3: Functional principle of a fiber Bragg grating.

A detailed insight and quantitative analysis of FBG's optical characteristics can be developed using coupled-mode theory [7]. The refractive index perturbation, responsible for the coupling between co- and counter-propagating modes in a uniform FBG (Fig. 22.3) can be written as

$$n(z) = n_0 + \Delta n \cos\left(\frac{2\pi z}{\Lambda}\right) \tag{22.4}$$

where n_0 is average refractive index, Δn is maximum amount of induced RI perturbation, and z is the distance along fiber axis. A quantitative expression to determine reflection spectrum of FBG having a sinusoidal varying RI perturbation as described by Eq. 22.4 can be obtained in terms of grating reflectivity (R) using coupled-mode theory; it is given by [8, 9]

$$R(L, \lambda) = \frac{\kappa^2 \sin h^2(\gamma L)}{\Delta k^2 \sin h^2(\gamma L) + \gamma^2 \cos h^2(\gamma L)} \tag{22.5}$$

where κ ($\approx \pi\eta\Delta n/\lambda$) is the coupling coefficient, η is the overlap integral for the transverse field distribution which is ≈ 1 for identical modes, β is the propagation constant, and $\gamma^2 = \kappa^2 - \Delta k^2$. As can be observed, grating reflectivity depends on its length and wavelength. Further, a maximum in the reflectivity is observed when

$$\Delta k = \beta - \frac{\pi}{\Lambda} = 0 \tag{22.6}$$

for which

$$R(L, \lambda)_{max} = \tanh^2 \kappa L \tag{22.7}$$

Equation 22.6 reduces to the Bragg condition (Eq. 22.3). For $\kappa L = 2$, $R_{max} = 0.93$. To achieve this characteristic reflectivity at 1550 nm with $\Delta n \approx 10^{-4}$, L must be ~5 mm. Another important parameter related to the uniform Bragg grating is its *bandwidth*, which is the wavelength spacing between first zeros on either side of the central peak and is given by [9]

$$\frac{\Delta\lambda}{\lambda_B} \approx \frac{\lambda_B}{n_{\text{eff}} L}\left[1+\left(\frac{(\Delta n)L}{\lambda_B}\right)^2\right]^{\frac{1}{2}} \tag{22.8}$$

For the FBG with specifications we used to calculate Λ and L, $\Delta\lambda$ comes out to be ≈ 1.7 nm.

In the next subsections, we shall focus on the optical spectrum and define the relevant optical characteristics of FBG.

22.4.3 Reflection spectrum of FBG

Figure 22.4 shows a simulated reflection spectrum of a fiber Bragg grating. The important characteristics of this spectrum, namely (1) peak reflectivity, (2) reflection (3 dB) bandwidth,

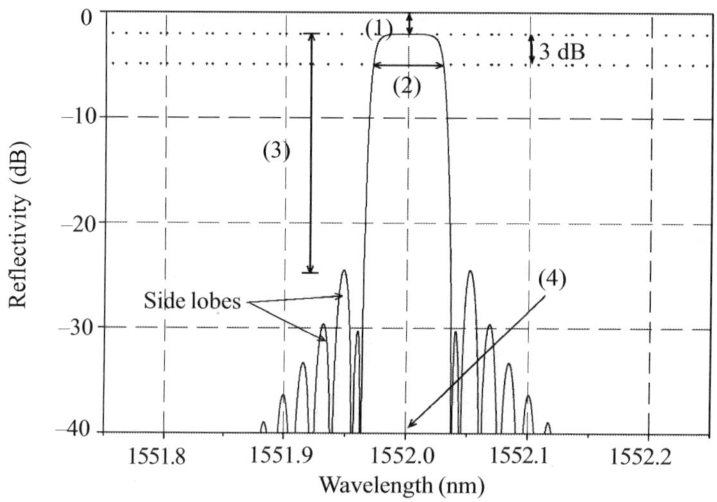

Fig. 22.4: Simulated reflection spectrum of fiber Bragg grating (with Gaussian apodization; apodization will be explained in the next chapter)

(3) amount of side lobe suppression or extinction ratio, and (4) center wavelength are depicted in this figure. The reflectivity is related to the strength of the FBG, which refers to the amount of light that can be reflected by a grating. In a given optical fiber link, FBG is incorporated either by splicing it or by using an optical circulator (will be explained in experimental section). Owing to the insertion from optical circulator or connector loss, the exact measurement of FBG's reflectivity in reflection mode is difficult and one must incorporate these losses. For example, if (1) is –1.6 dB, port 1–2 and port 2–3 losses of an optical circulator (see Fig. 22.6) are –0.55 dB and –0.35 dB respectively, and splice/connector loss is 0.21 dB, then the peak reflectivity would be calculated as $-1.6 \text{ dB} + 0.55 \text{ dB} + 0.35 \text{ dB} + 2 \times 0.21 \text{ dB} = -0.28$ dB. A more appropriate way is by measuring the transmission loss, which will be explained in the next subsection. The reflection 3 dB bandwidth defines the usable bandwidth of a FBG which reflects >50% of the incident

light. Spectral side lobe suppression or extinction ratio indicates the nearest wavelength for another WDM channel to be packed without inducing severe intra-channel cross talk. Thus, more signal wavelength channels can be closely placed if the extinction ratio of employed FBG is very high. This increases the bandwidth efficiency of a WDM system when the FBGs are chosen as multiplexers or demultiplexers. Center wavelength carries the grating periodicity information because a particular grating period is chosen based on the central wavelength satisfying Eq. (22.3). One must know the center wavelength in order to employ FBG as a wavelength filter.

22.4.4 Transmission spectrum of FBG

Reflectivity or strength of FBG can be precisely determined by characterizing the transmission spectrum of a grating. Accurate information about the reflectivity is important in order to determine the maximum amount of induced refractive index change (Δn) of grating. This information allows establishing a relation between the photosensitivity of the fiber, UV exposure time and the UV laser power. The exact reflectivity (R) of the FBG can be estimated through [8]

$$R = 1 - 10^{\left(-T_c/10\right)} \tag{22.9}$$

where T_c is the amount of the transmission loss at the center wavelength. Figure 22.5 depicts a typical transmission spectrum of a fiber Bragg grating. This spectrum has a transmission loss of approximately 21 dB at the center wavelength of the FBG. For this grating, R comes out to be

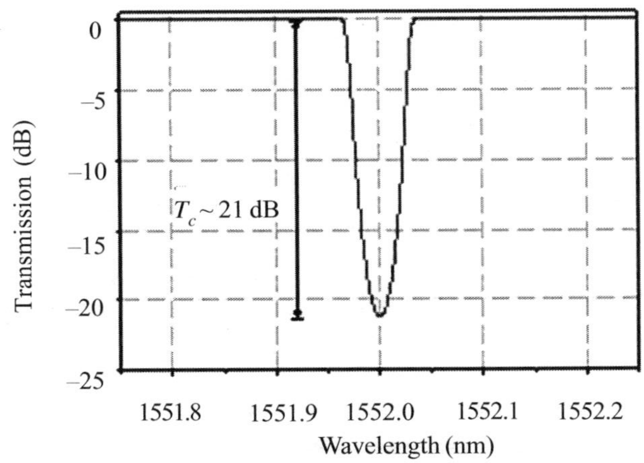

Fig. 22.5: Simulated transmission spectrum of fiber Bragg grating (with Gaussian apodization).

0.992 or 99.2%. The value of R can be further used for a rough estimation of maximum induced Δn as [8]

$$\Delta n \cong \frac{\lambda_c \tan h^{-1}(\sqrt{R})}{\pi L} \tag{22.10}$$

where $\boldsymbol{\lambda_c}$ is the center wavelength and L is the length of the FBG.

22.5 Procedure

Figure 22.6 illustrates the basic configuration for measuring the spectrum of a given FBG. Since FBG possesses the property of reflecting a specific wavelength as governed by its grating spacing, a combination of a broadband light source and a commercial optical spectrum analyzer (OSA) is used to perform the spectral characterization. To carry out the experiment, following procedures have to be followed step by step:

1. Clean all the fiber connectors of single-mode optical fiber patch cords (OFPC) using fiber cleaner.
2. Either an Erbium-doped ASE source or a white light source can be used as a broadband light source. The former has a better dynamic range than the latter. High dynamic range of >30 dB is important in order to resolve the details of reflection spectrum side lobes of the grating. Using a patch cord, connect the ASE source, which covers a broad wavelength range (usually from 1525 nm ~ 1570 nm), to the port "1" of an optical circulator.
3. Splice the given FBG to a patch cord and connect it to the port "2" of the circulator. Optical circulator directs the ASE signal from its port "1" to its port "2", and hence feeds the signal to the FBG.
4. Connect port "3" of the circulator to the "input" of an optical spectrum analyzer (OSA) using another patch cord. The reflected light from the FBG is directed to port "3" of the circulator. The measured spectrum at this port of the circulator determines the reflection spectrum of the FBG.
5. Light that is not reflected from the FBG is transported across it, forming the transmission spectrum of the grating. To measure this spectrum, OSA is connected to the port "2" of the circulator instead of port "3".

Note: A 3 dB (50 : 50) fused fiber coupler could also be used as an alternate of the expensive circulator for characterizing the FBG (see the inset of Fig. 22.6). In this case, port "A" of the coupler is connected to the broadband source through an isolator. Isolator prevents any back reflected light to reach to the source. FBG is spliced to the port "B".

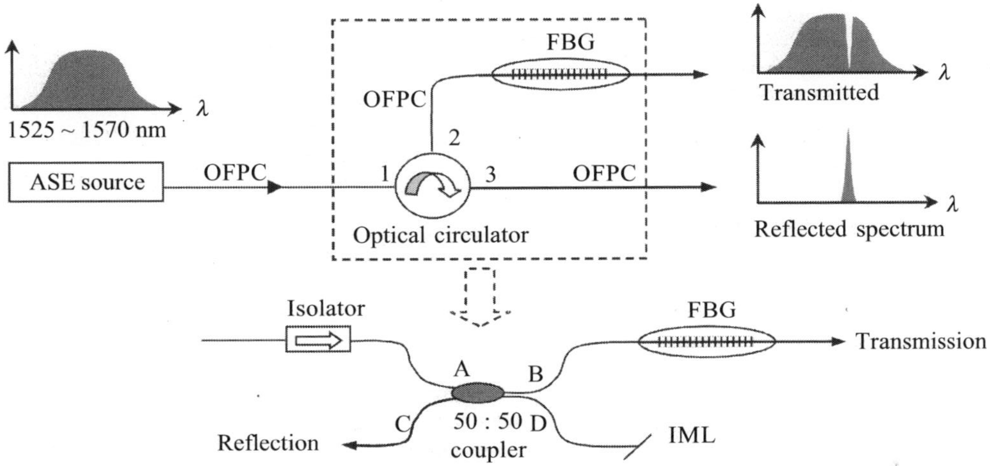

Fig. 22.6: Typical set-up for measuring reflection and transmission spectra of a fiber Bragg grating.

The OSA can be connected either to the port "B" (for transmission spectrum) or to the port "C" (for reflection spectrum). Index matching liquid (IML) is applied over a few centimeter length of the bare fiber near its end at the port "D". One of the main drawbacks of using coupler is that the measured reflection spectrum suffers an additional 6-dB loss as compared to the circulator method. However, this loss can be accounted by normalizing the input power.

22.6 Results

Figure 22.7 shows a typical reflection spectrum measured through OSA employing a 3 dB fused fiber coupler. Here, x-scale corresponds to 0.2 nm per division and y-scale corresponds to 2 dB per division. OSA use to have peak search option, which should be used to find peak reflectivity and corresponding wavelength. This wavelength is the Bragg wavelength λ_B, which is observed to be 1554.786 nm for the grating shown in Fig. 22.7. Peak power is observed to be ~ –30.5 dBm. As mentioned earlier, to find actual reflectivity, one must correct the observed reflectivity in terms of the losses of the system. Further, while measuring FBG spectra through an OSA, horizontal markers should be selected. One such marker "A" should be kept at the peak of the reflection spectrum (corresponding to the peak reflectivity), whereas another marker "B" should be brought to –3dB position with respect to the marker "A". Now vertical markers ("C" and "D") should be selected and brought to the intersection points of FBG spectrum and marker "B" (see Fig. 22.7). Vertical markers will read corresponding wavelengths on x-scale. Taking the difference of these wavelengths will define grating's 3 dB bandwidth. For the FBG shown in Fig. 22.7, 3 dB bandwidth was observed to be 0.26 nm.

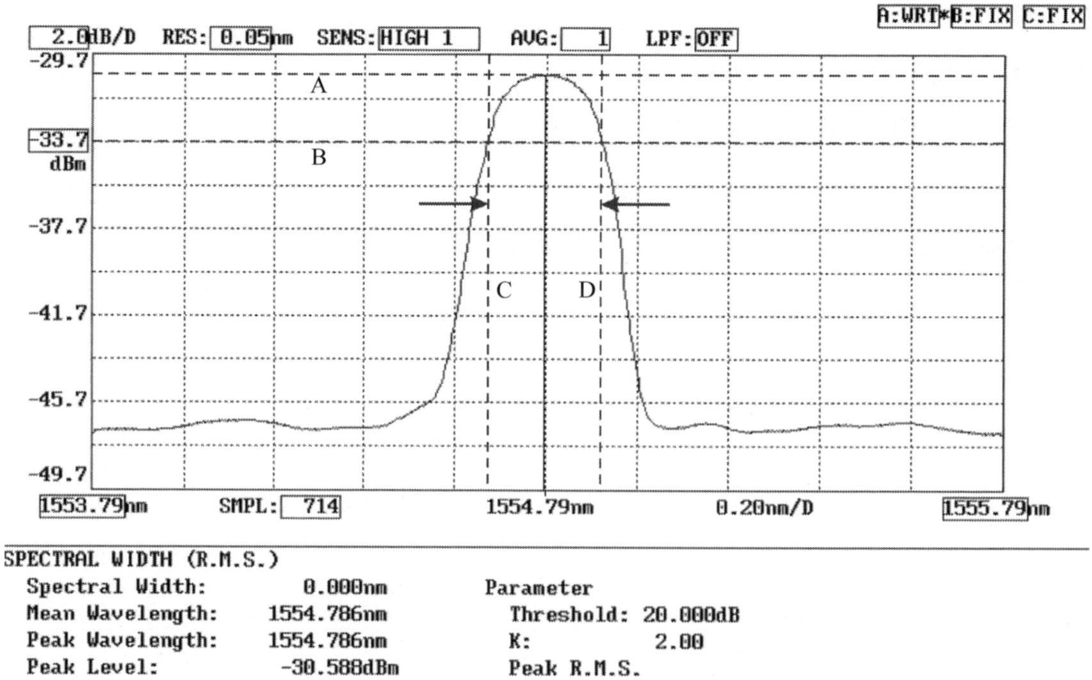

Fig. 22.7: Typical reflection spectrum of fiber Bragg grating measured through OSA. [Figure courtesy Prof. M.R. Shenoy, IIT Delhi. This FBG was fabricated at CGCRI, Kolkata.]

22.7 Comments

While carrying out the experiment, fiber end must be cleaned with fiber cleaner before connecting it to the source, detector etc. Un-cleaned connector may cause additional loss to the optical system. Further, the optical characteristics of a fiber Bragg grating are highly dependent on the temperature and strain variation along its length [10]. Care must be taken in this regard while carrying out the experiment.

Finally, it is worth to mention that the major drawback of the method employing OSA is the fact that the delay/dispersion characteristics of FBG cannot be measured. Further, this technique does not ensure acceptable quality of the grating. A grating appearing to be good in OSA spectrum (sharply peaked with no ripples) may not actually be so in practical terms. For FBG to be used for sensing applications, above mentioned discussions provided in this chapter should be sufficient, however, for its application in communication systems, one must characterize an FBG more extensively as discussed in the next chapter.

References

1. K.O. Hill, Y. Fujii, D.C. Johnson and B.S. Kawasaki, "Photosensitivity in optical waveguide: Application to reflection filter fabrication", *Appl. Phys. Lett.*, vol. 32, p. 647 (1978).
2. D.K.W. Lam and B.K. Garside, "Characterization of single-mode optical fiber filters", *Appl. Opt.*, vol. 20, p. 440 (1981).
3. B.P. Pal, "In-fiber grating: Evolution, optics, and applications in sensing", *Asian J. Phy.* (*Special issue on Fiber Optic Sensor*), vol. 12, p. 263 (2004).
4. D.L. Williams, B.J. Ainslie, R.J. Armitage, R. Kashyap and R.J. Campbell, "Enhanced photosensitivity in germanium doped silica fibers for future optical networks", *ECOC*'92, paper We B9-5, p. 425 (1992).
5. G. Mertz, W.W. Morey and W.H. Glenn, "Formation of Bragg grating in optical fibers by transverse holographic method", *Opt. Lett.*, vol. 14, p. 823 (1989).
6. K.O. Hill, B. Malo, F. Bilodeau, D.C. Johnson and J. Albert, "Bragg gratings fabricated in monomode photosensitive optical fiber by UV exposure through a phase mask", *Appl. Phys. Lett.*, vol. 62, p. 1035 (1993).
7. A. Ghatak and K. Thyagarajan, *Introduction to Fiber Optics*, Cambridge University Press, Cambridge (1998).
8. R. Kashyap, *Fiber Bragg Gratings,* Academic Press, USA (1999).
9. T. Erdogan, "Fiber Grating Spectra", *J. Lightwave Technol.*, vol. 15, p. 1277 (1997).
10. A. Othonos and K. Kalli, *Fiber Bragg Gratings*, Artech House, USA (1999).

23

Characterization of Fiber Bragg Grating - II

Sunil K. Khijwania and C.S. Goh

23.1 Introduction

Among other all-fiber components, one component which has left a great and wide-spread impact over the last decade on both the domains of optical fiber technology, namely – communication systems and sensing applications, is fiber Bragg grating (FBG). As discussed in the previous chapter, an intra-core periodic refractive index (RI) perturbation along the fiber length is termed as FBG. Most important point while incorporating an FBG in an optical network is to have proper and accurate information of its optical characteristics. For example, FBG with sharp reflection spectral edge and narrow bandwidth is ideal for filtering a modulated light. The sharp spectral roll-off also refrains the intra-channel cross talk from adjacent channels, especially in a high channel-count WDM system with channel spacing less than 50 GHz (0.4 nm). On the other hand, flattening out the amplified spontaneous emission (ASE) spectra of an erbium-doped fiber amplifier requires the FBG to have a transmission loss spectrum, which is inverse to that of the ASE. Another important application of FBG is dispersion compensation. An optical pulse gets broadened as it traverses the fiber communication link owing to the dispersion characteristics of the fiber. This broadening limits the information carrying capacity of the fiber optic system, and hence the pulse must be reshaped to its original configuration before being fed to an optical receiver. FBG used for this purpose usually possesses a nonlinear phase response across the useful bandwidth of the FBG, which is attributed to the chirp in the grating spacing along its length. Thus, it becomes very important to have an accurate optical analysis of FBG/chirped FBGs in terms of their reflection spectra, transmission spectra, phase and temporal responses in order to determine their employability in a desired application. We have introduced the basic concepts of fiber Bragg grating in the previous chapter along with a very simple experiment to determine reflectivity, bandwidth etc. of a given FBG. This technique is not very reliable if one needs to accurately analyze the optical response of a given fiber grating. As pointed out in Chapter 22, a grating though might appear to be good e.g. sharply peaked with no ripples on the OSA measurements; it may not necessarily be sufficient for practical purposes. Further, it is impossible to have an estimation of phase and temporal response of FBG with this method. This chapter introduces the basic concepts of *apodized* and *aperiodic* (chirped) FBGs, which are important for applications in optical communications systems. Further, an advanced experimental technique is discussed in order to accurately analyze the optical response of a given fiber gratings to study its reflection, transmission, phase and temporal characteristics.

23.2 Aim

To determine various optical characteristics of a given periodic/chirped fiber Bragg grating.

23.3 Apparatus

Tunable laser source, network analyzer, $LiNO_3$ modulator, lightwave detector, optical circulator/3 dB coupler, polarization controller, optical fiber splicer, patch cords, fiber connector cleaner and computer.

23.4 Theory

23.4.1 Apodized and chirped Fiber Bragg Grating

The salient features of uniform fiber Bragg gratings have been introduced in Chapter 22. One troublesome feature of such uniform gratings is that these typically exhibit side lobes of high amplitude in their reflection spectra. As an example, an experimentally observed accurate reflection spectrum of a uniform FBG (on the right hand side) along with the corresponding index modulation (on the left hand side) is depicted in Fig. 23.1(a). As can be observed, the reflection spectrum of a uniform FBG contains several side lobes on both sides of the central peak. As mentioned in the previous chapter, such side lobes are undesirable because these may cause cross talks between adjacent signal channels in a dense wavelength division multiplexing (DWDM) stream if the channels are closely spaced. The reason behind the occurrence of these side lobes is the formation of Fabry-Perot cavity owing to an abrupt RI change at $z = \pm\frac{L}{2}$, and its own resonance for the wavelengths outside the central peak. Here L is the grating length. Especially, in high channel count DWDM systems, it is necessary to suppress these side lobes efficiently in order to eliminate any possibility of cross talk between adjacent channels. Figure 23.1(b) shows a scheme to suppress the side lobes. Here, the intensity of UV beam is made non-uniform in order to taper down the index perturbation to zero at both ends of the FBG. The resulting refractive index modulation is depicted on the left hand side of Fig. 23.1(b).

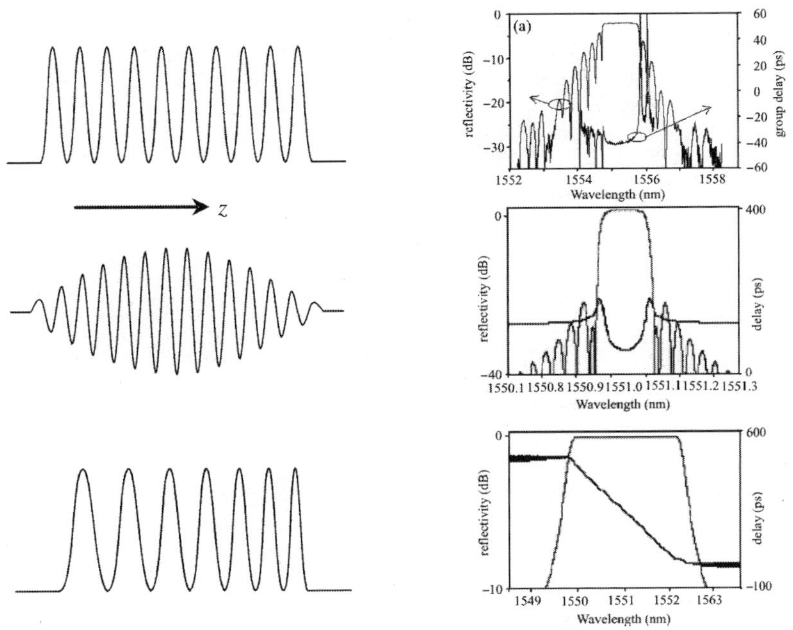

Fig. 23.1: Optical characteristics (in the right) and corresponding index modulation (in the left) of (a) uniform grating, (b) apodized grating, and (c) linearly chirped fiber Bragg grating.

This technique is termed as *apodization* and the resulting grating is termed as *apodized grating* [1–4]. The reflection spectrum of an *apodized* grating is depicted on the right side of Fig. 23.1(b). Compared to the optical characteristics of a uniform FBG [Fig. 23.1(a)], it can be observed that the side lobes are substantially suppressed in an *apodized* grating. It is worth to mention that all the definitions introduced in the section 22.4.3/22.4.4 of the last chapter for various optical characteristics of reflection/transmission spectrum remain the same for an apodized grating. However, a rough estimate of the maximum induced δn for an *apodized* grating could be made once R (reflectivity) is measured, and making use of the following relation [1, 3]:

$$\Delta n \cong \frac{\lambda_c \tan h^{-1}(\sqrt{R})}{aL\pi} \tag{23.1}$$

where λ_c is the center wavelength and L is the length of FBG. The factor "a" accounts for the apodization effect per unit length of the grating that is imposed on the FBG. The value of "a" depends on the choice of the apodization function (tanh, sine, raised sine, Blackman etc.). As an example, for an unapodized grating "a" is 1, whereas for strongest apodization such as Blackman function, "a" becomes 0.42 [3].

Another important class of fiber grating is termed as *chirped* (aperiodic) *fiber grating*. The term "*chirped*" is derived from its analogy to the continuously rising frequency that could be heard from a singing bird. Such gratings are realized by axially varying either Λ, or n_{eff} of the core, or both. One such resulting index modulation when the period is varied along the grating axis is depicted on the left hand side of Fig. 23.1(c). In this case, reflected wavelength also becomes a function of axial coordinate z [5]

$$\lambda_B(z) = 2\Lambda(z)n_{eff}(z) \tag{23.2}$$

The variation of $\Lambda(z)$ $n_{eff}(z)$ could either be linear or nonlinear, giving rise to linearly chirped FBG (LCFBG) or nonlinearly chirped FBG (NCFBG), respectively. In fact, the index modulation depicted in Fig. 23.1(c) corresponds to LCFBG. In the same figure, the optical characteristic of LCFBG is depicted on the right hand side. As can be observed, chirping increases the bandwidth. The most important application of these gratings is dispersion (both chromatic dispersion and dispersion slope), compensation and pulse compression. When a pulse is allowed to travel through this grating, its longer wavelength components are reflected from the front (corresponding to longer grating period), and shorter wavelength components from deep inside the grating (corresponding to shorter grating period) as shown in Fig. 23.2. As a result, shorter wavelengths return to the front end of the grating after a time delay as compared to the longer ones. Thus, the propagating pulse effectively experiences a large amount of group velocity dispersion. This is

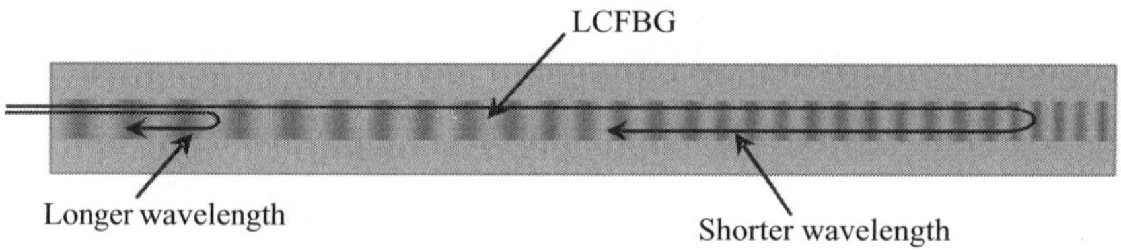

Fig. 23.2: Linearly chirped FBG's operation principle.

exactly opposite to what happens if the same pulse is allowed to propagate through a communication grade fiber in anomalous dispersion regime, where longer wavelengths travels slower than the shorter ones. If this pulse were now fed to LCFBG of appropriate length, then the dispersion induced pulse broadening would be completely compensated.

23.4.2 Phase response and dispersion of FBG

The phase response of FBG is very important for determining its role in the optical signal conditioning. For example, one requires a linear phase response across the grating bandwidth if the grating has to be used as a wavelength filter. However, the grating should possess a nonlinear phase response over its reflection bandwidth if it is to be used as a dispersion compensator. The nonlinear phase response is responsible for producing a wavelength dependent group delay or dispersion in the FBG. The *group delay* τ (or the *time delay* between various wavelength components of a pulse as it propagates through a FBG, which is usually given in units of picoseconds), and *dispersion D* (in ps/nm/km) of the reflected light for a given FBG can be determined from the phase (θ) of the amplitude reflection coefficient $\sqrt{R}$ (defined by Eq. 22.5 in Chapter 22). Group delay is given as [6, 7]

$$\tau(\lambda) = -\frac{\lambda^2}{2\pi c}\frac{d\theta}{d\lambda} \tag{23.3}$$

where c is the velocity of light in vacuum. Dispersion (D) of the FBG can then be estimated from the group delay as following:

$$D(\lambda) = \frac{d\tau}{d\lambda} \tag{23.4}$$

Figure 23.1 shows the delay for (a) uniform FBG, (b) apodized FBG, and (c) LCFBG in their respective optical spectra. Here, delay corresponds to the y-axis in the right side of the grating spectrum. For a uniform FBG, reflectivity and the delay are symmetric about the center wavelength. Apodization not only suppresses the side lobes of the reflection spectrum, but also reduces rapid oscillations of delay near the edges. Linear chirping produces a linear delay within the 3 dB bandwidth. It is this linear region, which is used to compensate group velocity dispersion incurred in optical fiber systems. To determine the amount of dispersion in the FBG, one needs to characterize its phase response. For this, a wavelength tunable laser source should be used to measure the phase shift incurred by the FBG at different wavelengths.

It can be concluded that the above-mentioned optical characteristics of FBG play a very crucial role towards its applications. Next section will introduce a measurement technique for the precise estimation of such characteristics of FBG.

23.5 Procedure

The amount of dispersion in the FBG can be determined by characterizing it's phase response using modulation phase shift (MPS) technique described in this section. This technique is not only important for dispersion measurement, but also for the exact reflection spectrum, giving accurate information about the ripple and side lobes of the grating. The set-up for measuring the dispersion of FBG using MPS technique is depicted in Fig. 23.3. A wavelength tunable CW laser with a wavelength λ_1 is modulated at a modulation frequency of f_M (Hz) (usually in hundred of MHz) from a network analyzer (NA) with a $LiNO_3$ modulator. The modulated light is then launched into the FBG (under test). The reflected light is converted back into the electrical signal through a lightwave detector and is compared with a reference signal (modulation frequency source). The difference between the detected signal and reference signal is the phase shift $\theta_{\lambda 1}$ for that wavelength, as shown in the inset (Fig. 23.3). The process is repeated again, by tuning the laser to another wavelength λ_2 such that $\lambda_2 = \lambda_1 + \Delta\lambda$, where $\Delta\lambda$ is the wavelength separation. The process continues until the laser scans the entire bandwidth of the FBG. This iterative process is controlled by a computer and the measured phase shifts $\theta_{\lambda 1}$, $\theta_{\lambda 2}$, … $\theta_{\lambda n}$ are saved in a file for further processing. The relative group delay ($\Delta\tau$) is then computed by the following formula:

$$\Delta\tau_{\lambda i+\frac{\Delta\lambda}{2}} = \frac{\theta_{\lambda(i+1)} - \theta_{\lambda i}}{2\pi f_M}, \text{ where } i = 1, 2, 3, \ldots n \tag{23.5}$$

The calculated $\Delta\tau_\lambda$ is then plotted as a function of wavelength. It can be observed from the above equation that the higher the modulation frequency f_M, the higher the resolution of the measurement system. Usually, the smallest phase difference that can be resolved is about 0.5° for a high-end phase meter. Therefore, for an f_M of 500 MHz, as small as 2.77 ps temporal resolution is possible.

As a more general case, the characterization results for chirped fiber Bragg grating would be presented in the next section. The same method could be applied to characterize a uniform FBG. Further, the main emphasis in the next section would be placed on measurements of phase and temporal characteristics of the grating. Once optical spectrum is obtained, other characteristics such as reflectivity, 3 dB bandwidth, extinction ratio etc. can be precisely estimated using the definition given in Chapter 22.

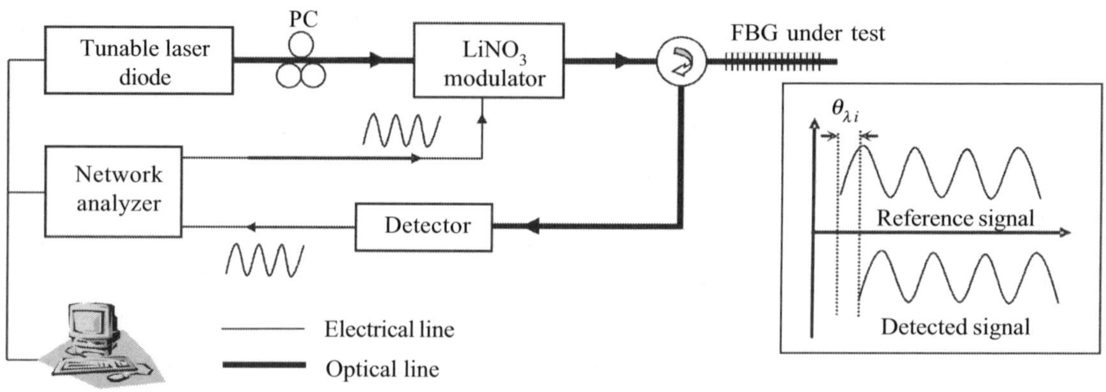

Fig. 23.3: Set-up for measuring various optical characteristics like reflectivity, phase, dispersion etc. of a fiber Bragg grating.

23.6 Results

23.6.1 Dispersion measurement for a linearly chirped FBG

Chirped fiber Bragg grating (CFBG) plays an extremely important role, especially in compensating dispersion of various orders in long haul communication systems [8]. Hence, the optical characteristics of such gratings become crucial in order to deploy them in a communication network. This section describes the characterization results for CFBG using MPS technique. Test FBG is replaced with CFBG in the experimental set-up. Figure 23.4 shows the optical characteristics of a 50 mm long CFBG. Here, the measurement is taken using a wavelength interval of 0.01 nm at modulation frequency of 625 MHz. Reflection spectrums corresponds to the *y*-axis on the left hand side with a reflection bandwidth of more than 6 nm. Further, measured relative group delay, which corresponds to the *y*-axis on the right hand side of the graph, exhibits linearity. The dispersion of FBG can be calculated by measuring the slope of this linear group delay within the grating bandwidth. By fitting the plot with a straight line, the dispersion of CFBG is estimated to be –55.2 ps/nm. This amount of normal dispersion can be used to compensate the anomalous dispersion of a 3.2 km long single-mode fiber (SMF), assuming the dispersion coefficient of SMF as 17 ps/nm/km.

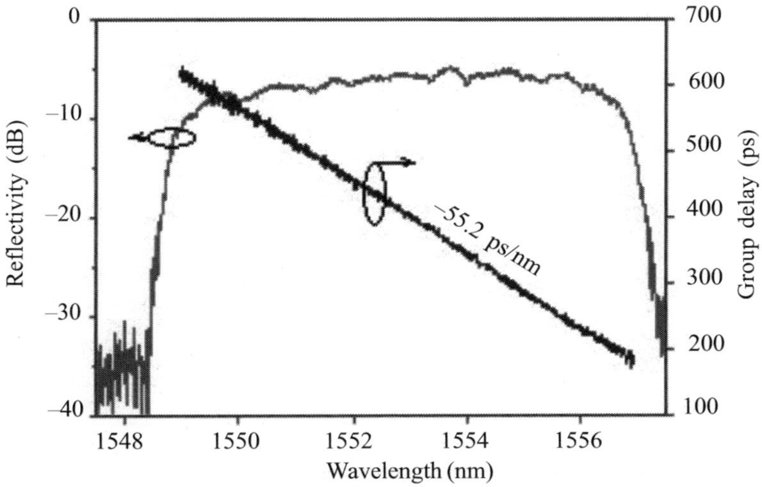

Fig. 23.4: Reflection and group delay of linearly chirped fiber Bragg grating

Another important characteristic of CFBG is the group delay ripple (GDR), which must be determined along with the grating dispersion. GDR is the deviation of the group delay from its

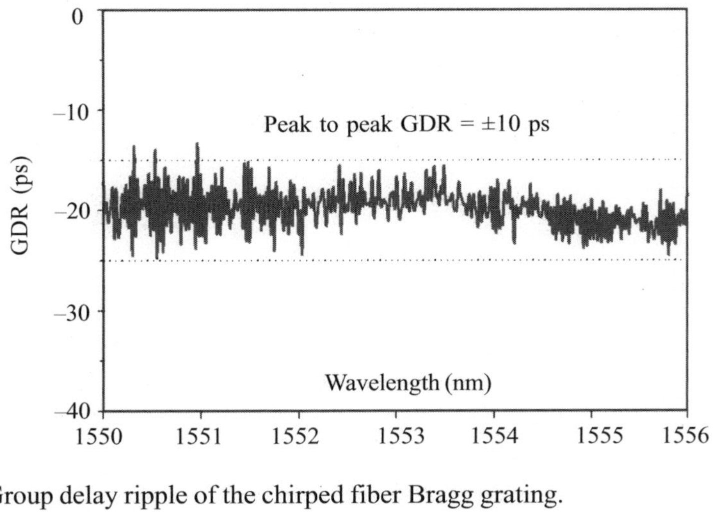

Fig. 23.5: Group delay ripple of the chirped fiber Bragg grating.

linearity. Excessive GDR may cause additional eye opening penalty (EOP) [9] to the dispersion compensated optical signal. GDR usually originates from the frozen imperfections while realizing FBG. Improper apodization also causes GDR. GDR is measured by taking the difference of observed group delay from a straight line fitting to the measured delay vs. wavelength curve. Figure 23.5 shows the GDR of the CFBG having reflection and delay characteristics as depicted in Fig. 23.4. The peak-to-peak GDR is observed to be around ±10 ps.

23.6.2 Dispersion slope measurement for a nonlinearly chirped FBG

The chromatic dispersion (D) of optical fiber degrades the transmission characteristics of optical communication systems. As discussed in Chapter 4, at a certain wavelength, (termed as *zero dispersion wavelength* λ_{ZD}), total chromatic dispersion becomes zero. Nevertheless, D can not be made zero for all frequencies within the pulse spectrum. Thus, even if the signal wavelength is adjusted to λ_{ZD}, in a high-speed WDM light-wave systems operating at 40 Gb/s and beyond, pulse gets broadened and distorted (in the form of trailing ripples) because of the finite dispersion slope $DS\ (= dD/d\lambda)$ [10]. Hence, in such high bit-rate systems, it becomes increasingly important to exactly compensate not only the chromatic dispersion, but also the dispersion slope of the fiber as they could impose a high dispersion penalty. *DS* compensation can be achieved by a nonlinearly chirped fiber Bragg grating (NCFBG) having nonlinear group delay across its bandwidth [8]. Therefore the characterization of NCFBG becomes important for such applications in order to have exact compensation. The nonlinear group delay along with the reflectivity of an NCFBG can be characterized once again by MPS technique. However, the amount of *DS* carried by the NCFBG has to be computed from the measured spectral characteristics.

Figure 23.6(a) shows the measured reflection spectrum (along y-axis on the LHS) and group delay profile (along y-axis on the RHS) of NCFBG. A nonlinear temperature distribution was applied across the FBG to realize NCFBG. The length of this NCFBG was 24 mm. The observed group delay of the NCFBG varies nonlinearly across its bandwidth. The slope of the fitted

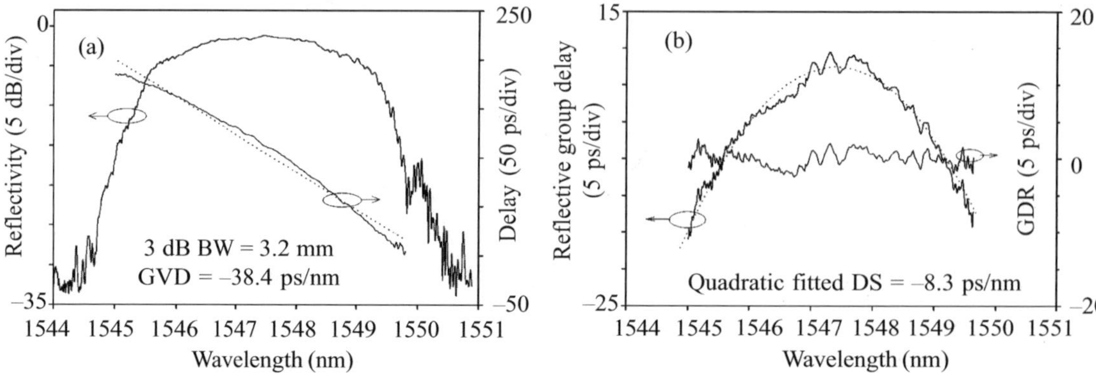

Fig. 23.6: Optical characteristics of an NCFBG, (a) reflection and delay, (b) quadratic delay and GDR.

straight line to the group delay within the grating bandwidth gives the residual dispersion (–38.4 ps/nm). Deviation of the actual nonlinear delay from the linear group delay (corresponding to the fitted straight line) gives a quadratic curve, which is a measure of induced quadratic delay [Fig. 23.6(b)]. In order to estimate the dispersion slope, a quadratic fitting to this curve is accomplished, as shown in Fig. 23.6(b). *DS* is directly obtained from this fitted curve. For the grating used in this case, a *DS* of –8.3 ps/nm^2/km was observed, which is sufficient to compensate the *DS* of a ~120 km long dispersion shifted fiber, having a *DS* of 0.07 ps/nm^2/km. Further, the peak-to-peak GDR [along *y*-axis in RHS of Fig. 23.6(b)] for this grating is observed to be around ±6 ps.

23.7 Comments

One must be careful in coupling the light into the fiber. It is better to take an estimate of optical power at each level of the set-up shown in Fig. 23.3. Further, the fiber end must be cleaned with fiber cleaner before connecting it to the source, detector etc. As pointed out in Chapter 22, characteristics of a fiber Bragg grating are highly dependent on the temperature and strain variation along its length [5]. Care must be taken in this regard while carrying out the experiment. With the accurately measured characteristics using MPS technique, the applicability and suitability of the FBG can be well determined for real world applications. Finally, the method to characterize a given FBG discussed in this chapter can be used for characterizing any FBG with any arbitrary profile (e.g., superstructure, tilted etc).

References

1. R. Kashyap, *Fiber Bragg Gratings,* Academic Press, USA (1999).
2. P.S. Cross and H. Kogelnik, "Sidelobe suppression in corrugated-wavelength filters", *Opt. Lett.*, vol. 1, p. 43 (1977).
3. D. Pastor, J. Campany, D. Ortega, V. Tatay and J. Marti, "Design of apodized linearly chirped grating for dispersion compensation", *J. Lightwave Technol.*, vol. 14, p. 2581 (1996).
4. C.R. Giles, "Lightwave applications of fiber Bragg gratings", *J. Lightwave Technol.*, vol. 15, p. 1391 (1997).
5. A. Othonos and K. Kalli, *Fiber Bragg Gratings*, Artech House, USA (1999).
6. T. Erdogan, "Fiber Grating Spectra", *J. Lightwave Technol.*, vol. 15, p. 1277 (1997).

7. A.K. Ghatak and K. Thyagarajan, *Optical Electronics*, Cambridge University Press, (1989).
8. S.K. Khijwania, C.S. Goh, S.Y. Set and K. Kikuchi, "Novel tunable dispersion slope compensator based on nonlinearly chirped fiber Bragg grating", *Optics Communications*, vol. 227, p 107 (2003).
9. G.P. Agrawal, *Fiber Optic Communicaion System*, Wiley-Interscience (1997).
10. G.P. Agrawal, *Nonlinear Fiber Optics*, Academic Press, USA (1989).

Index